Fermionic Functional Integrals and the Renormalization Group

Volume 16

CRM MONOGRAPH SERIES

Centre de Recherches Mathématiques
Université de Montréal

Fermionic Functional Integrals and the Renormalization Group

Joel Feldman
Horst Knörrer
Eugene Trubowitz

The Centre de Recherches Mathématiques (CRM) of the Université de Montréal was created in 1968 to promote research in pure and applied mathematics and related disciplines. Among its activities are special theme years, summer schools, workshops, postdoctoral programs, and publishing. The CRM is supported by the Université de Montréal, the Province of Québec (FCAR), and the Natural Sciences and Engineering Research Council of Canada. It is affiliated with the Institut des Sciences Mathématiques (ISM) of Montréal, whose constituent members are Concordia University, McGill University, the Université de Montréal, the Université du Québec à Montréal, and the Ecole Polytechnique. The CRM may be reached on the Web at www.crm.umontreal.ca.

American Mathematical Society
Providence, Rhode Island USA

The production of this volume was supported in part by the Fonds pour la Formation de Chercheurs et l'Aide à la Recherche (Fonds FCAR) and the Natural Sciences and Engineering Research Council of Canada (NSERC).

2000 *Mathematics Subject Classification.* Primary 82B28.

ABSTRACT. The Renormalization Group is the name given to a technique for analyzing the qualitative behaviour of a class of physical systems by iterating a map on the vector space of interactions for the class. In a typical nonrigorous application of this technique one assumes, based on one's physical intuition, that only a certain finite dimensional subspace (usually of dimension three or less) is important. These notes concern a technique for justifying this approximation in a broad class of fermionic models used in condensed matter and high energy physics.

These notes expand upon the Aisenstadt Lectures given by Joel Feldman at the Centre de recherches mathématiques, Université de Montréal in August 1999.

Library of Congress Cataloging-in-Publication Data

Feldman, Joel S., 1949–
Fermionic functional integrals and the renormalization group / Joel Feldman, Horst Knörrer, Eugene Trubowitz.
p. cm. — (CRM monograph series, ISSN 1065-8599 ; v. 16)
Includes bibliographical references.
ISBN 0-8218-2878-9 (alk. paper)
1. Integration, Functional. 2. Renormalization group. 3. Mathematical physics. I. Knörrer, Horst. II. Trubowitz, Eugene. III. Title. IV. Series.
QC20.7.F85 F45 2002
530.15′574—dc21 2002018586

∞ The paper used in this book is acid-free and falls within the guidelines
established to ensure permanence and durability.
This volume was submitted to the American Mathematical Society
in camera ready form by the Centre de Recherches Mathématiques.
Visit the AMS home page at URL: http://www.ams.org/

10 9 8 7 6 5 4 3 2 1 07 06 05 04 03 02

Contents

Preface

The Renormalization Group is the name given to a technique for analyzing the qualitative behavior of a class of physical systems by iterating a map on the vector space of interactions for the class. In a typical non-rigorous application of this technique one assumes, based on one's physical intuition, that only a certain finite dimensional subspace (usually of dimension three or less) is important. These notes concern a technique for justifying this approximation in a broad class of fermionic models used in condensed matter and high energy physics.

The first chapter provides the necessary mathematical background. Most of it is easy algebra—primarily the definition of Grassmann algebra and the definition and basic properties of a family of linear functionals on Grassmann algebras known as Grassmann Gaussian integrals. To make Section 1.1 really trivial, we consider only finite dimensional Grassmann algebras. A simple-minded method for handling the infinite dimensional case is presented in Appendix A. There is also one piece of analysis in Section 1.1—the Gram bound on Grassmann Gaussian integrals—and a brief discussion of how Grassmann integrals arise in quantum field theories.

The second chapter introduces an expansion that can be used to establish analytic control over the Grassmann integrals used in fermionic quantum field theory models, when the covariance (propagator) is "really nice." It is also used as one ingredient in a renormalization group procedure that controls the Grassmann integrals when the covariance is not so nice. To illustrate the latter, we look at the Gross-Neveu$_2$ model and at many-fermion models in two space dimensions.

CHAPTER 1

Fermionic Functional Integrals

This chapter just provides some mathematical background. Most of it is easy algebra—primarily the definition of Grassmann algebra and the definition and basic properties of a class of linear functionals on Grassmann algebras known as Grassmann Gaussian integrals. There is also one piece of analysis—the Gram bound on Grassmann Gaussian integrals—and a brief discussion of how Grassmann integrals arise in quantum field theories. To make this chapter really trivial, we consider only finite-dimensional Grassmann algebras. A simple-minded method for handling the infinite-dimensional case is presented in Appendix A.

1.1. Grassmann Algebras

DEFINITION 1.1 (Grassmann algebra with coefficients in $\mathbb{C}$). Let $\mathcal{V}$ be a finite-dimensional vector space over $\mathbb{C}$. The Grassmann algebra generated by $\mathcal{V}$ is

$$\bigwedge \mathcal{V} = \bigoplus_{n=0}^{\infty} \bigwedge^n \mathcal{V}$$

where $\bigwedge^0 \mathcal{V} = \mathbb{C}$ and $\bigwedge^n \mathcal{V}$ is the n-fold antisymmetric tensor product of $\mathcal{V}$ with itself. Thus, if $\{a_1, \ldots, a_D\}$ is a basis for $\mathcal{V}$, then $\bigwedge \mathcal{V}$ is a vector space with elements of the form

$$f(a) = \sum_{n=0}^{D} \sum_{1\le i_1<\cdots<i_n\le D} \beta_{i_1,\ldots,i_n} a_{i_1}\cdots a_{i_n}$$

with the coefficients $\beta_{i_i,\ldots,i_n} \in \mathbb{C}$. (In differential geometry, the traditional notation uses $a_{i_1}\wedge\cdots\wedge a_{i_n}$ in place of $a_{i_1}\cdots a_{i_n}$.) Addition and scalar multiplication is done componentwise. That is,

$$\begin{aligned}\alpha\Big(\sum_{n=0}^{D} \sum_{1\le i_1<\cdots<i_n\le D} \beta_{i_1,\ldots,i_n} a_{i_1}\cdots a_{i_n}\Big) + \gamma\Big(\sum_{n=0}^{D} \sum_{1\le i_1<\cdots<i_n\le D} \delta_{i_1,\ldots,i_n} a_{i_1}\cdots a_{i_n}\Big)\\ = \sum_{n=0}^{D} \sum_{1\le i_1<\cdots<i_n\le D} (\alpha\beta_{i_1,\ldots,i_n} + \gamma\delta_{i_1,\ldots,i_n}) a_{i_1}\cdots a_{i_n}.\end{aligned}$$

The multiplication in $\bigwedge \mathcal{V}$ is determined by distributivity and

$$(a_{i_1}\cdots a_{i_m})(a_{j_1}\cdots a_{j_n}) = \begin{cases} 0, & \{i_1,\ldots,i_m\}\cap\{j_1,\ldots,j_n\}\ne\varnothing;\\ \operatorname{sgn}(k)\, a_{k_1}\cdots a_{k_{m+n}}, & \text{otherwise,}\end{cases}$$

where $(k_1, \ldots, k_{m+n})$ is the permutation of $(i_1, \ldots, i_m, j_1 \ldots, j_n)$ with $k_1 < k_2 < \cdots < k_{m+n}$ and $\operatorname{sgn}(k)$ is the sign of that permutation. In particular

$$a_i a_j = -a_j a_i$$

and $a_i a_i = 0$ for all i. For example

$$\begin{aligned}[a_2 + 3a_3][a_1 + 2a_1 a_2] &= a_2 a_1 + 3a_3 a_1 + 2a_2 a_1 a_2 + 6a_3 a_1 a_2 \\ &= -a_1 a_2 - 3a_1 a_3 - 2a_1 a_2 a_2 - 6a_1 a_3 a_2 \\ &= -a_1 a_2 - 3a_1 a_3 + 6a_1 a_2 a_3.\end{aligned}$$

REMARK 1.2. If $\{a_1, \ldots, a_D\}$ is a basis for $\mathcal{V}$, then $\bigwedge \mathcal{V}$ has basis

$$\{a_{i_1} \cdots a_{i_n} \mid n \geq 0, 1 \leq i_1 < \cdots < i_n \leq D\}.$$

The dimension of $\bigwedge^n \mathcal{V}$ is $\binom{D}{n}$ and the dimension of $\bigwedge \mathcal{V}$ is $\sum_{n=0}^{D} \binom{D}{n} = 2^D$. Let

$$\mathcal{M}_n = \{(i_1, \ldots, i_n) \mid 1 \leq i_1, \ldots, i_n \leq D\}$$

be the set of all multi-indices of degree $n \geq 0$. Note that $i_1, i_2, \ldots,$ does not have to be in increasing order. For each $I \in \mathcal{M}_n$ set $a_I = a_{i_1} \cdots a_{i_n}$. Also set $a_\varnothing = 1$. Every element $f(a) \in \bigwedge \mathcal{V}$ has a unique representation

$$f(a) = \sum_{n=0}^{D} \sum_{I \in \mathcal{M}_n} \beta_I a_I$$

with the coefficients $\beta_I \in \mathbb{C}$ antisymmetric under permutations of the elements of I. For example, $a_1 a_2 = \frac{1}{2}[a_1 a_2 - a_2 a_1] = \beta_{(1,2)} a_{(1,2)} + \beta_{(2,1)} a_{(2,1)}$ with $\beta_{(1,2)} = -\beta_{(2,1)} = \frac{1}{2}$.

PROBLEM 1.1. Let $\mathcal{V}$ be a complex vector space of dimension D. Let $s \in \bigwedge \mathcal{V}$. Then s has a unique decomposition $s = s_0 + s_1$ with $s_0 \in \mathbb{C}$ and $s_1 \in \bigoplus_{n=1}^{D} \bigwedge^n \mathcal{V}$. Prove that, if $s_0 \neq 0$, then there is a unique $s' \in \bigwedge \mathcal{V}$ with $ss' = 1$ and a unique $s'' \in \bigwedge \mathcal{V}$ with $s''s = 1$ furthermore

$$s' = s'' = \frac{1}{s_0} + \sum_{n=1}^{D} (-1)^n \frac{s_1^n}{s_0^{n+1}}.$$

It will be convenient to generalize the concept of Grassmann algebra to allow for coefficients in more general algebras than $\mathbb{C}$. We shall allow the space of coefficients to be any superalgebra [**BS**]. Here are the definitions that do it.

DEFINITION 1.3 (Superalgebra). (i) A superalgebra is an associative algebra $\mathbb{S}$ with unit, denoted 1, together with a decomposition $\mathbb{S} = \mathbb{S}_+ \oplus \mathbb{S}_-$ such that $1 \in \mathbb{S}_+$,

$$\begin{aligned} &\mathbb{S}_+ \cdot \mathbb{S}_+ \subset \mathbb{S}_+, \qquad && \mathbb{S}_- \cdot \mathbb{S}_- \subset \mathbb{S}_+, \\ &\mathbb{S}_+ \cdot \mathbb{S}_- \subset \mathbb{S}_-, \qquad && \mathbb{S}_- \cdot \mathbb{S}_+ \subset \mathbb{S}_-, \end{aligned}$$

and

$$ss' = \begin{cases} s's, & \text{if } s \in \mathbb{S}_+ \text{ or } s' \in \mathbb{S}_+; \\ -s's, & \text{if } s, s' \in \mathbb{S}_-. \end{cases}$$

The elements of $\mathbb{S}_+$ are called even, the elements of $\mathbb{S}_-$ odd.

(ii) A graded superalgebra is an associative algebra $\mathbb{S}$ with unit, together with a decomposition $\mathbb{S} = \bigoplus_{m=0}^{\infty} \mathbb{S}_m$ such that $1 \in \mathbb{S}_0$, $\mathbb{S}_m \cdot \mathbb{S}_n \subset \mathbb{S}_{m+n}$ for all $m, n \geq 0$, such that the decomposition $\mathbb{S} = \mathbb{S}_+ \oplus \mathbb{S}_-$ with

$$\mathbb{S}_+ = \bigoplus_{m \text{ even}} \mathbb{S}_m, \qquad \mathbb{S}_- = \bigoplus_{m \text{ odd}} \mathbb{S}_m$$

gives $\mathbb{S}$ the structure of a superalgebra.

EXAMPLE 1.4. Let $\mathcal{V}$ be a complex vector space. The Grassmann algebra $\mathbb{S} = \bigwedge \mathcal{V} = \bigoplus_{m\geq 0} \bigwedge^m \mathcal{V}$ over $\mathcal{V}$ is a graded superalgebra. In this case, $\mathbb{S}_m = \bigwedge^m \mathcal{V}$, $\mathbb{S}_+ = \bigoplus_{m \text{ even}} \bigwedge^m \mathcal{V}$ and $\mathbb{S}_- = \bigoplus_{m \text{ odd}} \bigwedge^m \mathcal{V}$. In these notes, all of the superalgebras that we use will be Grassmann algebras.

DEFINITION 1.5 (Tensor product of two superalgebras). (i) Recall that the tensor product of two finite-dimensional vector spaces $\mathbb{S}$ and $\mathbb{T}$ is the vector space $\mathbb{S} \otimes \mathbb{T}$ constructed as follows. Consider the set of all formal sums $\sum_{i=1}^n s_i \otimes t_i$ with all of the s_i's in $\mathbb{S}$ and all of the t_i's in $\mathbb{T}$. Let $\equiv$ be the smallest equivalence relation on this set with

$$\begin{aligned} s\otimes t + s'\otimes t' &\equiv s'\otimes t' + s\otimes t \\ (s+s')\otimes t &\equiv s\otimes t + s'\otimes t \\ s\otimes(t+t') &\equiv s\otimes t + s\otimes t' \\ (zs)\otimes t &\equiv s\otimes(zt) \end{aligned}$$

for all $s, s' \in \mathbb{S}$, $t, t' \in \mathbb{T}$ and $z \in \mathbb{C}$. Then $\mathbb{S} \otimes \mathbb{T}$ is the set of all equivalence classes under this relation. Addition and scalar multiplication are defined in the obvious way.

(ii) Let $\mathbb{S}$ and $\mathbb{T}$ be superalgebras. We define multiplication in the tensor product $\mathbb{S} \otimes \mathbb{T}$ by

$$[s\otimes(t_+ + t_-)][(s_+ + s_-)\otimes t] = ss_+ \otimes t_+t + ss_+ \otimes t_-t + ss_- \otimes t_+t - ss_- \otimes t_-t$$

for $s \in \mathbb{S}$, $t \in \mathbb{T}$, $s_\pm \in \mathbb{S}_\pm$, $t_\pm \in \mathbb{T}_\pm$. This multiplication defines an algebra structure on $\mathbb{S}\otimes\mathbb{T}$. Setting $(\mathbb{S}\otimes\mathbb{T})_+ = (\mathbb{S}_+\otimes\mathbb{T}_+)\oplus(\mathbb{S}_-\otimes\mathbb{T}_-)$, $(\mathbb{S}\otimes\mathbb{T})_- = (\mathbb{S}_+\otimes\mathbb{T}_-)\oplus(\mathbb{S}_-\otimes\mathbb{T}_+)$ we get a superalgebra. If $\mathbb{S}$ and $\mathbb{T}$ are graded superalgebras then the decomposition $\mathbb{S} \otimes \mathbb{T} = \bigoplus_{m=0}^\infty (\mathbb{S} \otimes \mathbb{T})_m$ with

$$(\mathbb{S}\otimes\mathbb{T})_m = \bigoplus_{m_1+m_2=m} \mathbb{S}_{m_1} \otimes \mathbb{T}_{m_2}$$

gives $\mathbb{S} \otimes \mathbb{T}$ the structure of a graded superalgebra.

DEFINITION 1.6 (Grassmann algebra with coefficients in a superalgebra). Let $\mathcal{V}$ be a complex vector space and $\mathbb{S}$ be a superalgebra, the Grassmann algebra over $\mathcal{V}$ with coefficients in $\mathbb{S}$ is the superalgebra

$$\textstyle\bigwedge_{\mathbb{S}} \mathcal{V} = \mathbb{S} \otimes \bigwedge \mathcal{V}.$$

If $\mathbb{S}$ is a graded superalgebra, so is $\bigwedge_{\mathbb{S}} \mathcal{V}$.

REMARK 1.7. It is natural to identify $s \in \mathbb{S}$ with the element $s \otimes 1$ of $\bigwedge_{\mathbb{S}} \mathcal{V} = \mathbb{S} \otimes \bigwedge \mathcal{V}$ and to identify $a_{i_1} \cdots a_{i_n} \in \bigwedge \mathcal{V}$ with the element $1 \otimes a_{i_1} \cdots a_{i_n}$ of $\bigwedge_{\mathbb{S}} \mathcal{V}$. Under this identification

$$sa_{i_1}\cdots a_{i_n} = (s\otimes 1)(1\otimes a_{i_1}\cdots a_{i_n}) = s\otimes a_{i_1}\cdots a_{i_n}.$$

Every element of $\bigwedge_{\mathbb{S}} \mathcal{V}$ has a unique representation

$$\sum_{n=0}^{D} \sum_{1\leq i_1<\cdots<i_n\leq D} s_{i_1,\ldots,i_n} a_{i_1}\cdots a_{i_n}$$

with the coefficients $s_{i_1,\dots,i_n} \in \mathbb{S}$. Every element of $\bigwedge_{\mathbb{S}} \mathcal{V}$ also has a unique representation

$$\sum_{n=0}^{D} \sum_{I \in \mathcal{M}_n} s_I a_I$$

with the coefficients $s_I \in \mathbb{S}$ antisymmetric under permutation of the elements of I. If $\mathbb{S} = \bigwedge \mathcal{V}'$, with $\mathcal{V}'$ the complex vector space with basis $\{b_1, \dots, b_D\}$, then every element of $\bigwedge_{\mathbb{S}} \mathcal{V}$ has a unique representation

$$\sum_{n,m=0}^{D} \sum_{\substack{\mathrm{I} \in \mathcal{M}_n \\ \mathrm{J} \in \mathcal{M}_m}} \beta_{\mathrm{I},\mathrm{J}} b_{\mathrm{J}} a_{\mathrm{I}}$$

with the coefficients $\beta_{\mathrm{I},\mathrm{J}} \in \mathbb{C}$ separately antisymmetric under permutation of the elements of I and under permutation of the elements of J.

PROBLEM 1.2. Let $\mathcal{V}$ be a complex vector space of dimension D. Every element s of $\bigwedge \mathcal{V}$ has a unique decomposition $s = s_0 + s_1$ with $s_0 \in \mathbb{C}$ and $s_1 \in \bigoplus_{n=1}^{D} \bigwedge^n \mathcal{V}$. Define

$$e^s = e^{s_0} \left\{ \sum_{n=0}^{D} \frac{1}{n!} s_1^n \right\}.$$

Prove that if $s, t \in \bigwedge \mathcal{V}$ with $st = ts$, then, for all $n \in \mathbb{N}$,

$$(s+t)^n = \sum_{m=0}^{n} \binom{n}{m} s^m t^{n-m}$$

and

$$e^s e^t = e^t e^s = e^{s+t}$$

PROBLEM 1.3. Use the notation of Problem 1.2. Let, for each $\alpha \in \mathbb{R}$, $s(\alpha) \in \bigwedge \mathcal{V}$. Assume that $s(\alpha)$ is differentiable with respect to α (meaning that if we write $s(\alpha) = \sum_{n=0}^{D} \sum_{1 \le i_1 < \cdots < i_n \le D} s_{i_1,\dots,i_n}(\alpha) a_{i_1} \cdots a_{i_n}$, every coefficient $s_{i_1,\dots,i_n}(\alpha)$ is differentiable with respect to α) and that $s(\alpha)s(\beta) = s(\beta)s(\alpha)$ for all α and β. Prove that

$$\frac{d}{d\alpha} s(\alpha)^n = n s(\alpha)^{n-1} s'(\alpha)$$

and

$$\frac{d}{d\alpha} e^{s(\alpha)} = e^{s(\alpha)} s'(\alpha).$$

PROBLEM 1.4. Use the notation of Problem 1.2. If $s_0 > 0$, define

$$\ln s = \ln s_0 + \sum_{n=1}^{D} \frac{(-1)^{n-1}}{n} \left(\frac{s_1}{s_0} \right)^n$$

with $\ln s_0 \in \mathbb{R}$.

(a) Let, for each $\alpha \in \mathbb{R}$, $s(\alpha) \in \bigwedge \mathcal{V}$. Assume that $s(\alpha)$ is differentiable with respect to α, that $s(\alpha)s(\beta) = s(\beta)s(\alpha)$ for all α and β and that $s_0(\alpha) > 0$ for all α. Prove that

$$\frac{d}{d\alpha} \ln s(\alpha) = \frac{s'(\alpha)}{s(\alpha)}.$$

(b) Prove that if $s \in \bigwedge \mathcal{V}$ with $s_0 \in \mathbb{R}$, then

$$\ln e^s = s.$$

Prove that if $s \in \bigwedge \mathcal{V}$ with $s_0 > 0$, then

$$e^{\ln s} = s.$$

PROBLEM 1.5. Use the notation of Problems 1.2 and 1.4. Prove that if $s, t \in \bigwedge \mathcal{V}$ with $st = ts$ and $s_0, t_0 > 0$, then

$$\ln(st) = \ln s + \ln t.$$

PROBLEM 1.6. Generalize Problems 1.1–1.5 to $\bigwedge_{\mathbb{S}} \mathcal{V}$ with $\mathbb{S}$ a finite-dimensional graded superalgebra having $\mathbb{S}_0 = \mathbb{C}$.

PROBLEM 1.7. Let $\mathcal{V}$ be a complex vector space of dimension D. Let $s = s_0 + s_1 \in \bigwedge \mathcal{V}$ with $s_0 \in \mathbb{C}$ and $s_1 \in \bigoplus_{n=1}^{D} \bigwedge^n \mathcal{V}$. Let $f(z)$ be a complex valued function that is analytic in $|z| < r$. Prove that if $|s_0| < r$, then $\sum_{n=0}^{\infty} f^{(n)}(0) s^n / n!$ converges and

$$\sum_{n=0}^{\infty} \frac{1}{n!} f^{(n)}(0) s^n = \sum_{n=0}^{D} \frac{1}{n!} f^{(n)}(s_0) s_1^n.$$

1.2. Grassmann Integrals

Let $\mathcal{V}$ a finite-dimensional complex vector space and $\mathbb{S}$ a superalgebra. Given any *ordered* basis $\{a_1, \ldots, a_D\}$ for $\mathcal{V}$, the Grassmann integral $\int \cdot \, da_D \ldots da_1$ is defined to be the unique linear map from $\bigwedge_{\mathbb{S}} \mathcal{V}$ to $\mathbb{S}$ which is zero on $\bigoplus_{n=0}^{D-1} \bigwedge^n \mathcal{V}$ and obeys

$$\int a_1 \cdots a_D \, da_D \cdots da_1 = 1.$$

PROBLEM 1.8. Let $a_1, \ldots, a_D$ be an ordered basis for $\mathcal{V}$. Let $b_i = \sum_{j=1}^{D} M_{i,j} a_j$, $1 \le i \le D$, be another ordered basis for $\mathcal{V}$. Prove that

$$\int \cdot \, da_D \cdots da_1 = \det M \int \cdot \, db_D \cdots db_1.$$

In particular, if $b_i = a_{\sigma(i)}$ for some permutation $\sigma \in S_D$

$$\int \cdot \, da_D \cdots da_1 = \operatorname{sgn} \sigma \int \cdot \, db_D \cdots db_1.$$

EXAMPLE 1.8. Let $\mathcal{V}$ be a two-dimensional vector space with basis $\{a_1, a_2\}$ and $\mathcal{V}'$ be a second two-dimensional vector space with basis $\{b_1, b_2\}$. Set $\mathbb{S} = \bigwedge \mathcal{V}'$. Let $\lambda \in \mathbb{C} \setminus \{0\}$ and let S be the 2×2 skew symmetric matrix

$$S = \begin{bmatrix} 0 & -1/\lambda \\ 1/\lambda & 0 \end{bmatrix}.$$

Use S_{ij}^{-1} to denote the matrix element of

$$S^{-1} = \begin{bmatrix} 0 & \lambda \\ -\lambda & 0 \end{bmatrix}$$

in row i and column j. Then, using the definition of the exponential of Problem 1.2 and recalling that $a_1^2 = a_2^2 = 0$,

$$e^{-(\sum_{ij} a_i S_{ij}^{-1} a_j)/2} = e^{-\lambda[a_1 a_2 - a_2 a_1]/2} = e^{-\lambda a_1 a_2} = 1 - \lambda a_1 a_2$$

and

$$
\begin{aligned}
e^{\sum_i b_i a_i} e^{-(\sum_{ij} a_i S_{ij}^{-1} a_j)/2} &= e^{b_1 a_1 + b_2 a_2} e^{-\lambda a_1 a_2} \\
&= \Big\{1 + (b_1 a_1 + b_2 a_2) + \frac{1}{2}(b_1 a_1 + b_2 a_2)^2\Big\}\{1 - \lambda a_1 a_2\} \\
&= \{1 + b_1 a_1 + b_2 a_2 - b_1 b_2 a_1 a_2\}\{1 - \lambda a_1 a_2\} \\
&= 1 + b_1 a_1 + b_2 a_2 - (\lambda + b_1 b_2) a_1 a_2.
\end{aligned}
$$

Consequently, the integrals

$$
\int e^{-(\sum_{ij} a_i S_{ij}^{-1} a_j)/2}\, da_2\, da_1 = -\lambda
$$

$$
\int e^{\sum_i b_i a_i} e^{-(\sum_{ij} a_i S_{ij}^{-1} a_j)/2}\, da_2\, da_1 = -(\lambda + b_1 b_2)
$$

and their ratio is

$$
\frac{\int e^{\sum_i b_i a_i} e^{-(\sum_{ij} a_i S_{ij}^{-1} a_j)/2}\, da_2\, da_1}{\int e^{-(\sum_{ij} a_i S_{ij}^{-1} a_j)/2}\, da_2\, da_1} = -\frac{(\lambda + b_1 b_2)}{-\lambda} = 1 + \frac{1}{\lambda} b_1 b_2 = e^{-(\sum_{ij} b_i S_{ij} b_j)/2}.
$$

EXAMPLE 1.9. Let $\mathcal{V}$ be a $D = 2r$-dimensional vector space with basis $\{a_1, \ldots, a_D\}$ and $\mathcal{V}'$ be a second $2r$-dimensional vector space with basis $\{b_1, \ldots, b_D\}$. Set $\mathbb{S} = \bigwedge \mathcal{V}'$. Let $\lambda_1, \ldots, \lambda_r$ be nonzero complex numbers and let S be the $D \times D$ skew symmetric matrix

$$
S = \bigoplus_{m=1}^{r} \begin{bmatrix} 0 & -1/\lambda_m \\ 1/\lambda_m & 0. \end{bmatrix}
$$

All matrix elements of S are zero, except for r 2×2 blocks running down the diagonal. For example, if $r = 2$,

$$
S = \begin{bmatrix} 0 & -1/\lambda_1 & 0 & 0 \\ 1/\lambda_1 & 0 & 0 & 0 \\ 0 & 0 & 0 & -1/\lambda_2 \\ 0 & 0 & 1/\lambda_2 & 0. \end{bmatrix}
$$

Then, by the computations of Example 1.8,

$$
e^{-(\sum_{ij} a_i S_{ij}^{-1} a_j)/2} = \prod_{m=1}^{r} e^{-\lambda_m a_{2m-1} a_{2m}} = \prod_{m=1}^{r} \{1 - \lambda_m a_{2m-1} a_{2m}\}
$$

and

$$
\begin{aligned}
e^{\sum_i b_i a_i} e^{-(\sum_{ij} a_i S_{ij}^{-1} a_j)/2} &= \prod_{m=1}^{r} \{e^{b_{2m-1} a_{2m-1} + b_{2m} a_{2m}} e^{-\lambda_m a_{2m-1} a_{2m}}\} \\
&= \prod_{m=1}^{r} \{1 + b_{2m-1} a_{2m-1} + b_{2m} a_{2m} \\
&\qquad\qquad - (\lambda_m + b_{2m-1} b_{2m}) a_{2m-1} a_{2m}\}
\end{aligned}
$$

This time, the integrals

$$
\int e^{-(\sum_{ij} a_i S_{ij}^{-1} a_j)/2}\, da_D \cdots da_1 = \prod_{m=1}^{r} (-\lambda_m)
$$

$$
\int e^{\sum_i b_i a_i} e^{-(\sum_{ij} a_i S_{ij}^{-1} a_j)/2}\, da_D \cdots da_1 = \prod_{m=1}^{r} (-\lambda_m - b_{2m-1} b_{2m})
$$

and their ratio is

$$\frac{\int e^{\sum_i b_i a_i} e^{-(\sum_{ij} a_i S_{ij}^{-1} a_j)/2}\, da_D \cdots da_1}{\int e^{-(\sum_{ij} a_i S_{ij}^{-1} a_j)/2}\, da_D \cdots da_1} = \prod_{m=1}^{r} \left(1 + \frac{1}{\lambda} b_1 b_2\right) = e^{-(\sum_{ij} b_i S_{ij} b_j)/2}.$$

LEMMA 1.10. *Let $\mathcal{V}$ be a $D = 2r$-dimensional vector space with basis $\{a_1, \ldots, a_D\}$ and $\mathcal{V}'$ be a second D-dimensional vector space with basis $\{b_1, \ldots, b_D\}$. Set $\mathbb{S} = \bigwedge \mathcal{V}'$. Let S be a $D \times D$ invertible skew symmetric matrix. Then*

$$\frac{\int e^{\sum_i b_i a_i} e^{-(\sum_{ij} a_i S_{ij}^{-1} a_j)/2}\, da_D \cdots da_1}{\int e^{-(\sum_{ij} a_i S_{ij}^{-1} a_j)/2}\, da_D \cdots da_1} = e^{-(\sum_{ij} b_i S_{ij} b_j)/2}.$$

PROOF. Both sides of the claimed equation are rational, and hence meromorphic, functions of the matrix elements of S. So, by analytic continuation, it suffices to consider matrices S with real matrix elements. Because S is skew symmetric, $S_{jk} = -S_{kj}$ for all $1 \le j, k \le D$. Consequently, $\imath S$ is self-adjoint so that

- $\mathcal{V}$ has an orthonormal basis of eigenvectors of S,
- all eigenvalues of S are pure imaginary.

Because S is invertible, it cannot have zero as an eigenvalue. Because S has real matrix elements, $S\vec{v} = \mu\vec{v}$ implies $S\bar{\vec{v}} = \bar{\mu}\bar{\vec{v}}$ (with $^-$ designating "complex conjugate") so that

- the eigenvalues and eigenvectors of S come in complex conjugate pairs.

Call the eigenvalues of S, $\pm\imath 1/\lambda_1, \pm\imath 1/\lambda_2, \ldots, \pm\imath 1/\lambda_r$ and set

$$T = \bigoplus_{m=1}^{r} \begin{bmatrix} 0 & -1/\lambda_m \\ 1/\lambda_m & 0 \end{bmatrix}.$$

By Problem 1.9, below there exists a real orthogonal $D \times D$ matrix R such that $R^t S R = T$. Define

$$a_i' = \sum_{j=1}^{D} R_{ij}^t a_j, \qquad b_i' = \sum_{j=1}^{D} R_{ij}^t b_j.$$

Then, as R is orthogonal, $RR^t = \mathbb{1}$ so that $S = RTR^t$, $S^{-1} = RT^{-1}R^t$. Consequently,

$$\sum_i b_i' a_i' = \sum_{i,j,k} R_{ij}^t b_j R_{ik}^t a_k = \sum_{i,j,k} b_j R_{ji} R_{ik}^t a_k = \sum_{j,k} b_j \delta_{j,k} a_k = \sum_i b_i a_i$$

where $\delta_{j,k}$ is one if $j = k$ and zero otherwise. Similarly,

$$\sum_{ij} a_i S_{ij}^{-1} a_j = \sum_{ij} a_i' T_{ij}^{-1} a_j', \qquad \sum_{ij} b_i S_{ij} b_j = \sum_{ij} b_i' T_{ij} b_j'.$$

Hence, by Problem 1.8 and Example 1.9

$$\begin{aligned}
&\frac{\int e^{-\sum_i b_i a_i} e^{-(\sum_{ij} a_i S_{ij}^{-1} a_j)/2}\, da_D \cdots da_1}{\int e^{-(\sum_{ij} a_i S_{ij}^{-1} a_j)/2}\, da_D \cdots da_1} \\
&\qquad = \frac{\int e^{\sum_i b_i' a_i'} e^{-(\sum_{ij} a_i' T_{ij}^{-1} a_j')/2}\, da_D \cdots da_1}{\int e^{-(\sum_{ij} a_i' T_{ij}^{-1} a_j')/2}\, da_D \ldots da_1} \\
&\qquad = \frac{\int e^{\sum_i b_i' a_i'} e^{-(\sum_{ij} a_i' T_{ij}^{-1} a_j')/2}\, da_D' \cdots da_1'}{\int e^{-(\sum_{ij} a_1' T_{ij}^{-1} a_j')/2}\, da_D' \cdots da_1'}
\end{aligned}$$

$$= e^{-(\sum_{ij} b'_i T_{ij} b'_j)/2} = e^{-(\sum_{ij} b_i S_{ij} b_j)/2}. \qquad \square$$

PROBLEM 1.9. Let

- S be a matrix
- λ be a real number
- $\vec{v}_1$ and $\vec{v}_2$ be two mutually perpendicular, complex conjugate unit vectors
- $S\vec{v}_1 = \imath\lambda\vec{v}_1$ and $S\vec{v}_2 = \imath\lambda\vec{v}_2$.

Set

$$\vec{w}_1 = \frac{1}{\sqrt{2}\imath}(\vec{v}_1 - \vec{v}_2) \qquad \vec{w}_2 = \frac{1}{\sqrt{2}}(\vec{v}_1 + \vec{v}_2).$$

(a) Prove that

- $\vec{w}_1$ and $\vec{w}_2$ are two mutually perpendicular, real unit vectors
- $S\vec{w}_1 = \lambda\vec{w}_2$ and $S\vec{w}_2 = -\lambda\vec{w}_1$.

(b) Suppose, in addition, that S is a 2×2 matrix. Let R be the 2×2 matrix whose first column is $\vec{w}_1$ and whose second column is $\vec{w}_2$. Prove that R is a real orthogonal matrix and that $R^t S R = \left[\begin{smallmatrix} 0 & -\lambda \\ \lambda & 0 \end{smallmatrix}\right]$.

(c) Generalize to the case in which S is a $2r \times 2r$ matrix.

1.3. Differentiation and Integration by Parts

DEFINITION 1.11 (Left Derivative). Left derivatives are the linear maps from $\bigwedge \mathcal{V}$ to $\bigwedge \mathcal{V}$ that are determined as follows. For each $\ell = 1, \ldots, D$ and $\mathrm{I} \in \bigcup_{n=0}^{D} \mathcal{M}_n$ the left derivative $\partial/\partial a_\ell a_{\mathrm{I}}$ of the Grassmann monomial a_I is

$$\frac{\partial}{\partial a_\ell} a_{\mathrm{I}} = \begin{cases} 0, & \text{if } \ell \notin \mathrm{I}; \\ (-1)^{|\mathrm{J}|} a_{\mathrm{J}} a_{\mathrm{K}}, & \text{if } a_{\mathrm{I}} = a_{\mathrm{J}} a_\ell a_{\mathrm{K}}. \end{cases}$$

In the case of $\bigwedge_{\mathbb{S}} \mathcal{V}$, the left derivative is determined by linearity and

$$\frac{\partial}{\partial a_\ell} s a_{\mathrm{I}} = \begin{cases} 0, & \text{if } \ell \notin \mathrm{I}, \\ (-1)^{|\mathrm{J}|} s a_{\mathrm{J}} a_{\mathrm{K}}, & \text{if } a_{\mathrm{I}} = a_{\mathrm{J}} a_\ell a_{\mathrm{K}} \text{ and } s \in \mathbb{S}_+; \\ -(-1)^{|\mathrm{J}|} s a_{\mathrm{J}} a_{\mathrm{K}}, & \text{if } a_{\mathrm{I}} = a_{\mathrm{J}} a_\ell a_{\mathrm{K}} \text{ and } s \in \mathbb{S}_-. \end{cases}$$

EXAMPLE 1.12. For each $\mathrm{I} = \{i_1, \ldots i_n\}$ in $\mathcal{M}_n$,

$$\frac{\partial}{\partial a_{i_n}} \cdots \frac{\partial}{\partial a_{i_1}} a_{\mathrm{I}} = 1$$

$$\frac{\partial}{\partial a_{i_1}} \cdots \frac{\partial}{\partial a_{i_n}} a_I = (-1)^{|\mathrm{I}|(|\mathrm{I}|-1)/2}.$$

PROPOSITION 1.13 (Product Rule, Leibniz's Rule). *For all $k, \ell = 1, \ldots, D$, all $\mathrm{I}, \mathrm{J} \in \bigcup_{n=0}^{D} \mathcal{M}_n$ and any $f \in \bigwedge_{\mathbb{S}} \mathcal{V}$,*

(a) $\partial a_k / \partial a_\ell = \delta_{k,\ell}$

(b) $\partial(a_{\mathrm{I}} a_{\mathrm{J}})/\partial a_\ell = (\partial a_{\mathrm{I}}/\partial a_\ell) a_J + (-1)^{|\mathrm{I}|} a_I (\partial a_{\mathrm{J}}/\partial a_\ell)$

(c) $\int (\partial f / \partial a_\ell)\, da_D \cdots da_1 = 0$

(d) *The linear operators $\partial/\partial a_k$ and $\partial/\partial a_\ell$ anticommute. That is,*

$$\left(\frac{\partial}{\partial a_k}\frac{\partial}{\partial a_\ell} + \frac{\partial}{\partial a_\ell}\frac{\partial}{\partial a_k}\right) f = 0.$$

PROOF. Obvious, by direct calculation. $\square$

PROBLEM 1.10. Let $P(z) = \sum_{i\ge 0} c_i z^i$ be a power series with complex coefficients and infinite radius of convergence and let $f(a)$ be an *even* element of $\bigwedge_{\mathbb{S}} \mathcal{V}$. Show that

$$\frac{\partial}{\partial a_\ell} P\big(f(a)\big) = P'\big(f(a)\big)\Big(\frac{\partial}{\partial a_\ell} f(a)\Big).$$

EXAMPLE 1.14. Let $\mathcal{V}$ be a D-dimensional vector space with basis $\{a_1, \ldots, a_D\}$ and $\mathcal{V}'$ a second D-dimensional vector space with basis $\{b_1, \ldots, b_D\}$. Think of $e^{\sum_i b_i a_i}$ as an element of either $\bigwedge_{\bigwedge \mathcal{V}} \mathcal{V}')$ or $\bigwedge(\mathcal{V} \oplus \mathcal{V}')$. (Remark 1.7 provides a natural identification between $\bigwedge_{\bigwedge\mathcal{V}} \mathcal{V}'$, $\bigwedge_{\bigwedge\mathcal{V}'} \mathcal{V}$ and $\bigwedge(\mathcal{V}\oplus\mathcal{V}')$.) By the last problem with a_i replaced by b_i, $P(z) = e^z$ and $f(b) = \sum_i b_i a_i$,

$$\frac{\partial}{\partial b_\ell} e^{\sum_i b_i a_i} = e^{\sum_i b_i a_i} a_\ell.$$

Iterating

$$\begin{aligned}\frac{\partial}{\partial b_{i_1}} \cdots \frac{\partial}{\partial b_{i_n}} e^{\sum_i b_i a_i} &= \frac{\partial}{\partial b_{i_1}} \cdots \frac{\partial}{\partial b_{i_{n-1}}} e^{\sum_i b_i a_i} a_{i_n} \\ &= \frac{\partial}{\partial b_{i_1}} \cdots \frac{\partial}{\partial b_{i_{n-2}}} e^{\sum_i b_i a_i} a_{i_{n-1}} a_{i_n} = e^{\sum_i b_i a_i} a_{\mathrm{I}},\end{aligned}$$

where $\mathrm{I} = (i_1, \ldots, i_n) \in \mathcal{M}_n$. In particular,

$$a_{\mathrm{I}} = \frac{\partial}{\partial b_{i_1}} \cdots \frac{\partial}{\partial b_{i_n}} e^{\sum_i b_i a_i}\Big|_{b_1,\ldots,b_D=0}$$

where, as you no doubt guessed, $\sum_{\mathrm{I}} \beta_{\mathrm{I}} b_{\mathrm{I}}\big|_{b_1,\ldots,b_D=0}$ means $\beta_\varnothing$.

DEFINITION 1.15. Let $\mathcal{V}$ be a D-dimensional vector space and $\mathbb{S}$ a superalgebra. Let $S = (S_{ij})$ be a skew symmetric matrix of order D. The Grassmann Gaussian integral with covariance S is the linear map from $\bigwedge_{\mathbb{S}} \mathcal{V}$ to $\mathbb{S}$ determined by

$$\int e^{\sum_i b_i a_i}\, d\mu_S(a) = e^{-(\sum_{ij} b_i S_{ij} b_j)/2}.$$

REMARK 1.16. If the dimension D of $\mathcal{V}$ is even and if S is invertible, then, by Lemma 1.10,

$$\int f(a)\, d\mu_S(a) = \frac{\int f(a) e^{-(\sum_{ij} a_i S^{-1}_{ij} a_j)/2}\, da_D \cdots da_1}{\int e^{-(\sum_{ij} a_i S^{-1}_{ij} a_j)/2}\, da_D \cdots da_1}.$$

The Gaussian measure on $\mathbb{R}^D$ with covariance S (this time an invertible symmetric matrix) is

$$d\mu_S(\vec{x}) = \frac{e^{-(\sum_{ij} x_i S^{-1}_{ij} x_j)/2} d\vec{x}}{\int e^{-(\sum_{ij} x_i S^{-1}_{ij} x_j)/2}\, d\vec{x}}.$$

This is the motivation for the name "Grassmann Gaussian integral." Definition 1.15, however, makes sense even if D is odd, or S fails to be invertible. In particular, if S is the zero matrix, $\int f(a)\, d\mu_S(a) = f(0)$.

PROPOSITION 1.17 (Integration by Parts). *Let $S = (S_{ij})$ be a skew symmetric matrix of order D. Then, for each $k = 1, \ldots, D$,*

$$\int a_k f(a)\, d\mu_S(a) = \sum_{\ell=1}^{D} S_{k\ell} \int \frac{\partial}{\partial a_\ell} f(a)\, d\mu_S(a).$$

PROOF 1. This first argument, while instructive, is not complete. For it, we make the additional assumption that D is even. Since both sides are continuous in S, it suffices to consider S invertible. Furthermore, by linearity in f, it suffices to consider $f(a) = a_{\mathrm{I}}$, with $\mathrm{I} \in \mathcal{M}_n$. Then, by Proposition 1.13(c),

$$\begin{aligned}
0 &= \sum_{\ell=1}^{D} S_{k\ell} \int \left(\frac{\partial}{\partial a_\ell} a_{\mathrm{I}} e^{-(\sum a_i S^{-1}_{ij} a_j)/2} \right) da_D \cdots da_1 \\
&= \sum_{\ell=1}^{D} S_{k\ell} \int \left(\left(\frac{\partial}{\partial a_\ell} \right) - (-1)^{|I|} a_{\mathrm{I}} \sum_{m=1}^{D} S^{-1}_{\ell m} a_m \right) e^{-(\sum a_i S^{-1}_{ij} a_j)/2} \, da_D \cdots da_1 \\
&= \sum_{\ell=1}^{D} S_{k\ell} \int \left(\frac{\partial}{\partial a_\ell} a_{\mathrm{I}} \right) e^{-(\sum a_i S^{-1}_{ij} a_j)/2} \, da_D \cdots da_1 \\
&\qquad\qquad - \int (-1)^{|I|} a_{\mathrm{I}} a_k e^{-(\sum a_i S^{-1}_{ij} a_j)/2} \, da_D \ldots da_1 \\
&= \sum_{\ell=1}^{D} S_{k\ell} \int \left(\frac{\partial}{\partial a_\ell} a_{\mathrm{I}} \right) e^{-(\sum a_i S^{-1}_{ij} a_j)/2} \, da_D \ldots da_1 \\
&\qquad\qquad - \int a_k a_{\mathrm{I}} e^{-(\sum a_i S^{-1}_{ij} a_j)/2} \, da_D \cdots da_1 .
\end{aligned}$$

Consequently,

$$\int a_k a_{\mathrm{I}} \, d\mu_S(a) = \sum_{\ell=1}^{n} S_{k\ell} \int \frac{\partial}{\partial a_\ell} a_{\mathrm{I}} \, d\mu_S(a). \qquad \square$$

PROOF 2. By linearity, it suffices to consider $f(a) = a_{\mathrm{I}}$ with $\mathrm{I} = \{i_1, \ldots, i_n\} \in \mathcal{M}_n$. Then

$$\begin{aligned}
\int a_k a_{\mathrm{I}} \, d\mu_S(a) &= \int \frac{\partial}{\partial b_k} \frac{\partial}{\partial b_{i_1}} \cdots \frac{\partial}{\partial b_{i_n}} e^{\sum_i b_i a_i} \Big|_{b_1,\ldots,b_D=0} d\mu_S(a) \\
&= \frac{\partial}{\partial b_k} \frac{\partial}{\partial b_{i_1}} \cdots \frac{\partial}{\partial b_{i_n}} \int e^{\sum_i b_i a_i} \, d\mu_S(a) \Big|_{b_1,\ldots,b_D=0} \\
&= \frac{\partial}{\partial b_k} \frac{\partial}{\partial b_{i_1}} \cdots \frac{\partial}{\partial b_{i_n}} e^{-(\sum_{i\ell} b_i S_{i\ell} b_\ell)/2} \Big|_{b_1,\ldots,b_D=0} \\
&= (-1)^n \frac{\partial}{\partial b_{i_1}} \cdots \frac{\partial}{\partial b_{i_n}} \frac{\partial}{\partial b_k} e^{-(\sum_{i\ell} b_i S_{i\ell} b_\ell)/2} \Big|_{b_1,\ldots,b_D=0} \\
&\quad - (-1)^n \sum_\ell S_{k\ell} \frac{\partial}{\partial b_{i_1}} \cdots \frac{\partial}{\partial b_{i_n}} b_\ell e^{-(\sum_{ij} b_i S_{ij} b_j)/2} \Big|_{b_1,\ldots,b_D=0} \\
&= -(-1)^n \sum_\ell S_{k\ell} \frac{\partial}{\partial b_{i_1}} \cdots \frac{\partial}{\partial b_{i_n}} b_\ell \int e^{\sum_i b_i a_i} \, d\mu_S(a) \Big|_{b_1,\ldots,b_D=0} \\
&= (-1)^n \sum_\ell S_{k\ell} \frac{\partial}{\partial b_{i_1}} \cdots \frac{\partial}{\partial b_{i_n}} \int \frac{\partial}{\partial a_\ell} e^{\sum_i b_i a_i} \, d\mu_S(a) \Big|_{b_1,\ldots,b_D=0} \\
&= \sum_\ell S_{k\ell} \int \frac{\partial}{\partial a_\ell} \frac{\partial}{\partial b_{i_1}} \cdots \frac{\partial}{\partial b_{i_n}} e^{\sum_i b_i a_i} \Big|_{b_1,\ldots,b_D=0} d\mu_S(a) \\
&= \sum_\ell S_{k\ell} \int \frac{\partial}{\partial a_\ell} a_{\mathrm{I}} \, d\mu_S(a). \qquad \square
\end{aligned}$$

1.4. Grassmann Gaussian Integrals

We now look more closely at the values of

$$\int a_{i_1} \cdots a_{i_n} \, d\mu_S(a) = \frac{\partial}{\partial b_{i_1}} \cdots \frac{\partial}{\partial b_{i_n}} e^{-(\sum_{i\ell} b_i S_{i\ell} b_\ell)/2} \Big|_{b_1,\ldots,b_D=0}.$$

Obviously,

$$\int 1 \, d\mu_S(a) = 1$$

and, as $\sum_{i,\ell} b_i S_{i\ell} b_\ell$ is even, $\partial/\partial b_{i_1} \cdots \partial/\partial b_{i_n} e^{-(\sum_{i\ell} b_i S_{i\ell} b_\ell)/2}$ has the same parity as n and

$$\int a_{i_1} \cdots a_{i_n} \, d\mu_S(a) = 0 \quad \text{if } n \text{ is odd}$$

To get a little practice with integration by parts (Proposition 1.17) we use it to evaluate $\int a_{i_1} \ldots a_{i_n} \, d\mu_S(a)$ when $n = 2$

$$\begin{aligned} \int a_{i_1} a_{i_2} \, d\mu_S &= \sum_{m=1}^{D} S_{i_1 m} \int \Big(\frac{\partial}{\partial a_m} a_{i_2} \Big) \, d\mu_S \\ &= \sum_{m=1}^{D} S_{i_1 m} \int \delta_{m,i_2} \, d\mu_S \\ &= S_{i_1 i_2} \end{aligned}$$

and $n = 4$

$$\begin{aligned} \int a_{i_1} a_{i_2} a_{i_3} a_{i_4} \, d\mu_S &= \sum_{m=1}^{D} S_{i_1 m} \int \Big(\frac{\partial}{\partial a_m} a_{i_2} a_{i_3} a_{i_4} \Big) \, d\mu_S \\ &= \int (S_{i_1 i_2} a_{i_3} a_{i_4} - S_{i_1 i_3} a_{i_2} a_{i_4} + S_{i_1 i_4} a_{i_2} a_{i_3}) \, d\mu_S \\ &= S_{i_1 i_2} S_{i_3 i_4} - S_{i_1 i_3} S_{i_2 i_4} + S_{i_1 i_4} S_{i_2 i_3}. \end{aligned}$$

Observe that each term is a sign times $S_{i_{\pi(1)} i_{\pi(2)}} S_{i_{\pi(3)} i_{\pi(4)}}$ for some permutation π of $(1, 2, 3, 4)$. The sign is always the sign of the permutation. Furthermore, for each odd j, $\pi(j)$ is smaller than all of $\pi(j+1), \pi(j+2), \ldots,$ (because each time we applied $\int a_{i_\ell} \cdots d\mu_S = \sum_m S_{i_\ell m} \int \partial/\partial a_m \cdots \, d\mu_S$, we always used the smallest available ℓ). From this you would probably guess that

$$\int a_{i_1} \cdots a_{i_n} \, d\mu_S(a) = \sum_\pi \operatorname{sgn} \pi \, S_{i_{\pi(1)} i_{\pi(2)}} \cdots S_{i_{\pi(n-1)} i_{\pi(n)}} \tag{1.1}$$

where the sum is over all permutations π of $1, 2, \ldots, n$ that obey

$$\pi(1) < \pi(3) < \cdots < \pi(n-1) \quad \text{and} \quad \pi(k) < \pi(k+1) \text{ for all } k = 1, 3, 5, \ldots, n-1.$$

Another way to write the same thing, while avoiding the ugly constraints, is

$$\int a_{i_1} \cdots a_{i_n} \, d\mu_S(a) = \frac{1}{2^{n/2} (n/2)!} \sum_\pi \operatorname{sgn} \pi \, S_{i_{\pi(1)} i_{\pi(2)}} \cdots S_{i_{\pi(n-1)} i_{\pi(n)}}$$

where now the sum is over all permutations π of $1, 2, \ldots, n$. The right-hand side is precisely the definition of the Pfaffian of the $n \times n$ matrix whose (ℓ, m) matrix element is $S_{i_\ell i_m}$.

Pfaffians are closely related to determinants. The following proposition gives their main properties. This proposition is proven in Appendix B.

PROPOSITION 1.18. *Let $S = (S_{ij})$ be a skew symmetric matrix of even order $n = 2m$.*

(a) *For all $1 \le k \ne \ell \le n$, let $M_{k\ell}$ be the matrix obtained from S by deleting rows k and ℓ and columns k and ℓ. Then,*

$$\mathrm{Pf}(S) = \sum_{\ell=1}^{n} \mathrm{sgn}(k-\ell)(-1)^{k+\ell} S_{k\ell}\,\mathrm{Pf}(M_{k\ell}).$$

In particular,

$$\mathrm{Pf}(S) = \sum_{\ell=2}^{n} (-1)^{\ell} S_{1\ell}\,\mathrm{Pf}(M_{1\ell}).$$

(b) *If*

$$S = \begin{pmatrix} \mathbf{0} & C \\ -C^t & \mathbf{0} \end{pmatrix}$$

where C is a complex $m \times m$ matrix. Then.

$$\mathrm{Pf}(S) = (-1)^{m(m-1)/2} \det(C).$$

(c) *For any $n \times n$ matrix B,*

$$\mathrm{Pf}(B^t S B) = \det(B)\,\mathrm{Pf}(S).$$

(d) $\mathrm{Pf}(S)^2 = \det(S)$.

Using these properties of Pfaffians (in particular the "expansion along the first row" of Proposition 1.18(a)) we can easily verify that our conjecture (1.1) was correct.

PROPOSITION 1.19. *For all even $n \ge 2$ and all $1 \le i_1, \ldots, i_n \le D$,*

$$\int a_{i_1} \cdots a_{i_n}\, d\mu_S(a) = \mathrm{Pf}[S_{i_k i_\ell}]_{1 \le k,\ell \le n}$$

PROOF. The proof is by induction on n. The statement of the proposition has already been verified for $n = 2$. Integrating by parts,

$$\begin{aligned} \int a_{i_1} \cdots a_{i_n}\, d\mu_S(a) &= \sum_{\ell=1}^{n} S_{i_1 \ell} \int \Big(\frac{\partial}{\partial a_\ell} a_{i_2} \cdots a_{i_n}\Big)\, d\mu_S(a) \\ &= \sum_{j=2}^{n} (-1)^j S_{i_1 i_j} \int a_{i_2} \ldots a_{i_{j-1}} a_{i_{j+1}} \cdots a_{i_n}\, d\mu_S(a). \end{aligned}$$

Our induction hypothesis and Proposition 1.18(a) now imply

$$\int a_{i_1} \cdots a_{i_n}\, d\mu_S(a) = \mathrm{Pf}[S_{i_k i_\ell}]_{1 \le k,\ell \le n}.$$ □

Hence we may use the following as an alternative to Definition 1.15.

DEFINITION 1.20. Let $\mathbb{S}$ be a superalgebra, $\mathcal{V}$ be a finite-dimensional complex vector space and S a skew symmetric bilinear form on $\mathcal{V}$. Then the Grassmann Gaussian integral on $\bigwedge_{\mathbb{S}} \mathcal{V}$ with covariance S is the $\mathcal{S}$-linear map

$$f(a) \in \textstyle\bigwedge_{\mathbb{S}} \mathcal{V} \longmapsto \int f(a)\, d\mu_S(a) \in \mathbb{S}$$

that is determined as follows. Choose a basis $\{a_i \mid 1 \le i \le D\}$ for $\mathcal{V}$. Then

$$\int a_{i_1} a_{i_2} \cdots a_{i_n} \, d\mu_S(a) = \mathrm{Pf}[S_{i_k i_\ell}]_{1 \le k,\ell \le n}$$

where $S_{ij} = S(a_i, a_j)$.

PROPOSITION 1.21. *Let S and T be skew symmetric matrices of order D. Then*

$$\int f(a) \, d\mu_{S+T}(a) = \int \left[\int f(a+b) \, d\mu_S(a) \right] d\mu_T(b).$$

PROOF. Let $\mathcal{V}$ be a D-dimensional vector space with basis $\{a_1, \ldots, a_D\}$. Let $\mathcal{V}'$ and $\mathcal{V}''$ be two more copies of $\mathcal{V}$ with bases $\{b_1, \ldots, b_D\}$ and $\{c_1, \ldots, c_D\}$ respectively. It suffices to consider $f(a) = e^{\sum_i c_i a_i}$. Viewing $f(a)$ as an element of $\bigwedge_{\bigwedge_S \mathcal{V}''} \mathcal{V}$,

$$\int f(a) \, d\mu_{S+T}(a) = \int e^{\sum_i c_i a_i} \, d\mu_{S+T}(a) = e^{-(\sum_{ij} c_i (S_{ij}+T_{ij}) c_j)/2}.$$

Viewing $f(a+b) = e^{\sum_i c_i (a_i + b_i)}$ as an element of $\bigwedge_{\bigwedge_S (\mathcal{V}'+\mathcal{V}'')} \mathcal{V}$,

$$\begin{aligned}
\int f(a+b) \, d\mu_S(a) &= \int e^{\sum_i c_i (a_i+b_i)} \, d\mu_S(a) \\
&= e^{\sum_i c_i b_i} \int e^{\sum_i c_i a_i} \, d\mu_S(a) \\
&= e^{\sum_i c_i b_i} e^{-(\sum_{ij} c_i S_{ij} c_j)/2}.
\end{aligned}$$

Now viewing $e^{\sum_i c_i b_i} e^{\sum_{ij} c_i S_{ij} c_j}$ as an element of $\bigwedge_{\bigwedge_S \mathcal{V}''} \mathcal{V}'$

$$\begin{aligned}
\int \left[\int f(a+b) \, d\mu_S(a) \right] d\mu_T(b) &= e^{-(\sum_{ij} c_i S_{ij} c_j)/2} \int e^{\sum_i c_i b_i} \, d\mu_T(b) \\
&= e^{-(\sum_{ij} c_i S_{ij} c_j)/2} e^{-(\sum_{ij} c_i T_{ij} c_j)/2} \\
&= e^{-(\sum_{ij} c_i (S_{ij}+T_{ij}) c_j)/2}
\end{aligned}$$

as desired. □

PROBLEM 1.11. Let $\mathcal{V}$ and $\mathcal{V}'$ be vector spaces with basis $\{a_1, \ldots, a_D\}$ and $\{b_1, \ldots, b_{D'}\}$ respectively. Let S and T be $D \times D$ and $D' \times D'$ skew symmetric matrices. Prove that

$$\int \left[\int f(a,b) \, d\mu_S(a) \right] d\mu_T(b) = \int \left[\int f(a,b) \, d\mu_T(b) \right] d\mu_S(a).$$

PROBLEM 1.12. Let $\mathcal{V}$ be a D-dimensional vector space with basis $\{a_1, \ldots, a_D\}$ and $\mathcal{V}'$ be a second copy of $\mathcal{V}$ with basis $\{c_1, \ldots, c_D\}$. Let S be a $D \times D$ skew symmetric matrix. Prove that

$$\int e^{\sum_i c_i a_i} \int f(a) \, d\mu_S(a) = e^{-(\sum_{ij} c_i S_{ij} c_j)/2} \int f(a - Sc) \, d\mu_S(a).$$

Here $(Sc)_i = \sum_j S_{ij} c_j$.

1.5. Grassmann Integrals and Fermionic Quantum Field Theories

In a quantum mechanical model, the set of possible states of the system form (the rays in) a Hilbert space $\mathcal{H}$ and the time evolution of the system is determined by a self-adjoint operator H on $\mathcal{H}$, called the Hamiltonian. We shall denote by Ω a ground state of H (eigenvector of H of lowest eigenvalue). In a quantum field theory, there is additional structure. There is a special family, $\{\varphi(\mathbf{x},\sigma) \mid \mathbf{x} \in \mathbb{R}^d, \sigma \in \mathfrak{S}\}$ of operators on $\mathcal{H}$, called annihilation operators. Here d is the dimension of space and $\mathfrak{S}$ is a finite set. You should think of $\varphi(\mathbf{x},\sigma)$ as destroying a particle of spin σ at $\mathbf{x}$. The adjoints, $\{\varphi^\dagger(\mathbf{x},\sigma) \mid \mathbf{x} \in \mathbb{R}^d, \sigma \in \mathfrak{S}\}$, of these operators are called creation operators. You should think of $\varphi^\dagger(\mathbf{x},\sigma)$ as creating a particle of spin σ at $\mathbf{x}$. All states in $\mathcal{H}$ can be expressed as linear combinations of products of annihilation and creation operators applied to Ω. The time evolved annihilation and creation operators are

$$e^{iHt}\varphi(\mathbf{x},\sigma)e^{-iHt}, \qquad\qquad e^{iHt}\varphi^\dagger(\mathbf{x},\sigma)e^{-iHt}.$$

If you are primarily interested in thermodynamic quantities, you should analytically continue these operators to imaginary $t = -i\tau$

$$e^{H\tau}\varphi(\mathbf{x},\sigma)e^{-H\tau}, \qquad\qquad e^{H\tau}\varphi^\dagger(\mathbf{x},\sigma)e^{-H\tau}$$

because the density matrix for temperature T is $e^{-\beta H}$ where $\beta = 1/(kT)$. The imaginary time operators (or rather, various inner products constructed from them) are also easier to deal with mathematically rigorously than the corresponding real time inner products. It has turned out tactically advantageous to attack the real time operators by first concentrating on imaginary time and then analytically continuing back.

If you are interested in grand canonical ensembles (thermodynamics in which you adjust the average energy of the system through β and the average density of the system through the chemical potential μ) you replace the Hamiltonian H by $K = H - \mu N$, where N is the number operator and μ is the chemical potential. This brings us to

$$\begin{aligned} \varphi(\tau,\mathbf{x},\sigma) &= e^{K\tau}\varphi(\mathbf{x},\sigma)e^{-K\tau}, \\ \bar\varphi(\tau,\mathbf{x},\sigma) &= e^{K\tau}\varphi^\dagger(\mathbf{x},\sigma)e^{-K\tau}. \end{aligned} \tag{1.2}$$

Note that $\bar\varphi(\tau,\mathbf{x},\sigma)$ is neither the complex conjugate, nor adjoint, of $\varphi(\tau,\mathbf{x},\sigma)$.

In any quantum mechanical model, the quantities you measure (called observables) are represented by operators on $\mathcal{H}$. The expected value of the observable $\mathcal{O}$ when the system is in state Ω is $\langle \Omega, \mathcal{O}\Omega\rangle$, where $\langle\cdot,\cdot\rangle$ is the inner product on $\mathcal{H}$. In a quantum field theory, all expected values are determined by inner products of the form

$$\Big\langle \Omega, \mathbb{T}\prod_{\ell=1}^{p} \overset{(-)}{\varphi}(\tau_\ell,\mathbf{x}_\ell,\sigma_\ell)\Omega\Big\rangle.$$

Here the $\overset{(-)}{\varphi}$ signifies that both φ and $\bar\varphi$ may appear in the product. The symbol $\mathbb{T}$ designates the "time ordering" operator, defined (for fermionic models) by

$$\mathbb{T}\overset{(-)}{\varphi}(\tau_1,\mathbf{x}_1,\sigma_1)\cdots\overset{(-)}{\varphi}(\tau_p,\mathbf{x}_p,\sigma_p) = \operatorname{sgn}\pi\,\overset{(-)}{\varphi}(\tau_{\pi(1)},\mathbf{x}_{\pi(1)},\sigma_{\pi(1)})\cdots\overset{(-)}{\varphi}(\tau_{\pi(p)},\mathbf{x}_{\pi(p)},\sigma_{\pi(p)}),$$

where π is a permutation that obeys $\tau_{\pi(i)} \geq \tau_{\pi(i+1)}$ for all $1 \leq i \leq p$. There is also a tie breaking rule to deal with the case when $\tau_i = \tau_j$ for some $i \neq j$, but it doesn't interest us here. Observe that

$$\begin{aligned}\bar{\varphi}(\tau_{\pi(1)},\mathbf{x}_{\pi(1)},\sigma_{\pi(1)})\cdots\bar{\varphi}(\tau_{\pi(p)},\mathbf{x}_{\pi(p)},\sigma_{\pi(p)})&\\ = e^{K\tau_{\pi(1)}}\varphi^{(\dagger)}(\mathbf{x}_{\pi(1)},\sigma_{\pi(1)})e^{K(\tau_{\pi(2)}-\tau_{\pi(1)})}\varphi^{(\dagger)}\cdots&\\ \times e^{K(\tau_{\pi(p)}-\tau_{\pi(p-1)})}\varphi^{(\dagger)}(\mathbf{x}_{\pi(p)},\sigma_{\pi(p)})e^{-K\tau_{\pi(p)}}.&\end{aligned}$$

The time ordering is chosen so that, when you substitute in (1.2), and exploit the fact that Ω is an eigenvector of K, every $e^{K(\tau_{\pi(i)}-\tau_{\pi(i-1)})}$ that appears in $\langle\Omega,\mathbb{T}\prod_{\ell=1}^{p}\bar{\varphi}(\tau_\ell,\mathbf{x}_\ell,\sigma_\ell)\Omega\rangle$ has $\tau_{\pi(i)}-\tau_{\pi(i-1)}\le 0$. Time ordering is introduced to ensure that the inner product is well-defined: the operator K is bounded from below, but not from above. Thanks to the $\operatorname{sgn}\pi$ in the definition of $\mathbb{T}$, $\langle\Omega,\mathbb{T}\prod_{\ell=1}^{p}\bar{\varphi}(\tau_\ell,\mathbf{x}_\ell,\sigma_\ell)\Omega\rangle$ is antisymmetric under permutations of the $\bar{\varphi}(\tau_\ell,\mathbf{x}_\ell,\sigma_\ell)$'s.

Now comes the connection to Grassmann integrals. Let $x=(\tau,\mathbf{x})$. Please forget, for the time being, that x does not run over a finite set. Let $\mathcal{V}$ be a vector space with basis $\{\psi_{x,\sigma},\bar\psi_{\mathbf{x},\sigma}\}$. Note, once again, that $\bar\psi_{x,\sigma}$ is *NOT* the complex conjugate of $\psi_{x,\sigma}$. It is just another vector that is totally independent of $\psi_{x,\sigma}$. Then, formally, it turns out that

$$(-1)^p\Big\langle\Omega,\mathbb{T}\prod_{\ell=1}^{p}\bar{\varphi}(x_\ell,\sigma_\ell)\Omega\Big\rangle=\frac{\int\prod_{\ell=1}^{p}\bar{\psi}_{x_\ell,\sigma_\ell}e^{\mathcal{A}(\psi,\bar\psi)}\prod_{x,\sigma}d\psi_{x,\sigma}\,d\bar\psi_{x,\sigma}}{\int e^{\mathcal{A}(\psi,\bar\psi)}\prod_{x,\sigma}d\psi_{x,\sigma}\,d\bar\psi_{x,\sigma}}$$

The exponent $\mathcal{A}(\psi,\bar\psi)$ is called the action and is determined by the Hamiltonian H. A typical action of interest is that corresponding to a gas of fermions (e.g., electrons), of strictly positive density, interacting through a two-body potential $u(\mathbf{x}-\mathbf{y})$. It is

$$\begin{aligned}\mathcal{A}(\psi,\bar\psi)=-\sum_{\sigma\in\mathfrak{S}}\int\frac{d^{d+1}k}{(2\pi)^{d+1}}\Big(ik_0-\Big(\frac{\mathbf{k}^2}{2m}-\mu\Big)\Big)\bar\psi_{k,\sigma}\psi_{k,\sigma}&\\ -\frac{\lambda}{2}\sum_{\sigma,\sigma'\in\mathfrak{S}}\int\prod_{i=1}^{4}\frac{d^{d+1}k_i}{(2\pi)^{d+1}}(2\pi)^{d+1}\delta(k_1+k_2-k_3-k_4)\bar\psi_{k_1,\sigma}\psi_{k_3,\sigma}\hat{u}(\mathbf{k}_1-\mathbf{k}_3)\bar\psi_{k_2,\sigma'}\psi_{k_4,\sigma'}.&\end{aligned}$$

Here $\psi_{x,\sigma}$ is the Fourier transform of $\psi_{x,\sigma}$. The zero component k_0 of k is the dual variable to τ and is thought of as an energy; the final d components $\mathbf{k}$ are the dual variables to $\mathbf{x}$ and are thought of as momenta. So, in $\mathcal{A}$, $\mathbf{k}^2/2m$ is the kinetic energy of a particle and the delta function $\delta(k_1+k_2-k_3-k_4)$ enforces conservation of energy/momentum. As above, μ is the chemical potential, which controls the density of the gas, and $\hat{u}$ is the Fourier transform of the two-body potential. More generally, when the fermion gas is subject to a periodic potential due to a crystal lattice, the quadratic term in the action is replaced by

$$-\sum_{\sigma\in\mathfrak{S}}\int\frac{d^{d+1}k}{(2\pi)^{d+1}}\big(ik_0-e(\mathbf{k})\big)\bar\psi_{k,\sigma}\psi_{k,\sigma}$$

where $e(\mathbf{k})$ is the dispersion relation minus the chemical potential μ.

I know that quite a few people are squeamish about dealing with infinite-dimensional Grassmann algebras and integrals. Infinite-dimensional Grassmann algebras, per se, are no big deal. See Appendix A. It is true that the Grassmann "Cartesian measure" $\prod_{x,\sigma}d\psi_{x,\sigma}d\bar\psi_{x,\sigma}$ does not make much sense when the dimension is infinite. But this problem is easily dealt with: combine the quadratic part of the action $\mathcal{A}$ with $\prod_{x,\sigma} d\psi_{x,\sigma}\, d\bar\psi_{x,\sigma}$ to form a Gaussian integral. Formally

$$(1.3)\quad (-1)^p\Big\langle \Omega, \mathbb{T}\prod_{\ell=1}^{p} \overset{(-)}{\psi}(\tau_\ell, \mathbf{x}_\ell, \sigma_\ell\Omega\Big\rangle = \frac{\int \prod_{\ell=1}^{p} \overset{(-)}{\psi}_{x_\ell,\sigma_\ell} e^{\mathcal{A}(\psi,\bar\psi)} \prod_{x,\sigma} d\psi_{x,\sigma}\, d\bar\psi_{x,\sigma}}{\int e^{\mathcal{A}(\psi,\bar\psi)} \prod_{x,\sigma} d\psi_{x,\sigma}\, d\bar\psi_{x,\sigma}}$$
$$= \frac{\int \prod_{\ell=1}^{p} \overset{(-)}{\psi}_{x_\ell,\sigma_\ell} e^{W(\psi,\bar\psi)}\, d\mu_S(\psi,\bar\psi)}{\int e^{W(\psi,\bar\psi)}\, d\mu_S(\psi,\bar\psi)}$$

where

$$W(\psi,\bar\psi) = -\frac{\lambda}{2}\sum_{\sigma,\sigma'\in\mathfrak{S}} \int \prod_{i=1}^{4} \frac{d^{d+1}k_i}{(2\pi)^{d+1}} (2\pi)^{d+1}\delta(k_1+k_2-k_3-k_4)$$
$$\times\, \bar\psi_{k_1,\sigma}\psi_{k_3,\sigma}\hat u(\mathbf{k}_1-\mathbf{k}_3)\bar\psi_{k_2,\sigma'}\psi_{k_4,\sigma'}$$

and $\int \cdot\, d\mu_S(\psi,\bar\psi)$ is the Grassmann Gaussian integral with covariance determined by

$$(1.4)\quad \begin{aligned}
&\int \psi_{k,\sigma}\psi_{p,\sigma'}\, d\mu_S(\psi,\bar\psi) = 0\\
&\int \bar\psi_{k,\sigma}\bar\psi_{p,\sigma'}\, d\mu_S(\psi,\bar\psi) = 0\\
&\int \psi_{k,\sigma}\bar\psi_{p,\sigma'}\, d\mu_S(\psi,\bar\psi) = \frac{\delta_{\sigma,\sigma'}}{ik_0 - e(\mathbf{k})}(2\pi)^{d+1}\delta(k-p)\\
&\int \bar\psi_{k,\sigma}\psi_{p,\sigma'}\, d\mu_S(\psi,\bar\psi) = -\frac{\delta_{\sigma,\sigma'}}{ik_0 - e(\mathbf{k})}(2\pi)^{d+1}\delta(k-p).
\end{aligned}$$

PROBLEM 1.13. Let $\mathcal{V}$ be a complex vector space with even dimension $D = 2r$ and basis $\{\psi_1,\dots,\psi_r,\bar\psi_1,\dots,\bar\psi_r\}$. Again, $\bar\psi_i$ need not be the complex conjugate of ψ_i. Let A be an $r\times r$ matrix and $\int \cdot\, d\mu_A(\psi,\bar\psi)$ the Grassmann Gaussian integral obeying

$$\int \psi_i\psi_j\, d\mu_A(\psi,\bar\psi) = \int \bar\psi_i\bar\psi_j\, d\mu_A(\psi,\bar\psi) = 0, \qquad \int \psi_i\bar\psi_j\, d\mu_A(\psi,\bar\psi) = A_{ij}.$$

(a) Prove

$$\int \psi_{i_n}\cdots\psi_{i_1}\bar\psi_{j_1}\cdots\bar\psi_{j_m}\, d\mu_A(\psi,\bar\psi)$$
$$= \sum_{\ell=1}^{m}(-1)^{\ell+1}A_{i_1 j_\ell}\int \psi_{i_n}\cdots\psi_{i_2}\bar\psi_{j_1}\cdots\not{\bar\psi}_{j_\ell}\cdots\bar\psi_{j_m}\, d\mu_A(\psi,\bar\psi).$$

Here, the $\not{\bar\psi}_{j_\ell}$ signifies that the factor $\bar\psi_{j_\ell}$ is omitted from the integrand.

(b) Prove that, if $n\neq m$,

$$\int \psi_{i_n}\cdots\psi_{i_1}\bar\psi_{j_1}\cdots\bar\psi_{j_m}\, d\mu_A(\psi,\bar\psi) = 0.$$

(c) Prove that

$$\int \psi_{i_n}\cdots\psi_{i_1}\bar\psi_{j_1}\cdots\bar\psi_{j_n}\, d\mu_A(\psi,\bar\psi) = \det\left[A_{i_k j_\ell}\right]_{1\le k,\ell\le n}.$$

(d) Let $\mathcal{V}'$ be a second copy of $\mathcal{V}$ with basis $\{\zeta_1, \ldots, \zeta_r, \bar{\zeta}_1, \ldots, \bar{\zeta}_r\}$. View $e^{\sum_i (\bar{\zeta}_i \psi_i + \bar{\psi}_i \zeta_i)}$ as an element of $\bigwedge_{\bigwedge \mathcal{V}'} \mathcal{V}$. Prove that

$$\int e^{\sum_i (\bar{\zeta}_i \psi_i + \bar{\psi}_i \zeta_i)} \, d\mu_A(\psi, \bar{\psi}) = e^{\sum_{i,j} \bar{\zeta}_i A_{ij} \zeta_j}.$$

To save writing, lets rename the generators of the Grassmann algebra from $\{\psi_{k,\sigma}, \bar{\psi}_{k,\sigma}\}$ to $\{a_i\}$. In this new notation, the right-hand side of (1.3) becomes

$$G(i_1, \ldots, i_p) = \frac{1}{Z} \int \prod_{\ell=1}^{p} a_{i_\ell} e^{W(a)} \, d\mu_S(a), \quad \text{where } Z = \int e^{W(a)} \, d\mu_S(a).$$

These are called the Euclidean Green's functions or Schwinger functions. They determine all expectation values in the model. Let $\{c_i\}$ be the basis of a second copy of the vector space with basis $\{a_i\}$. Then all expectation values in the model are also determined by

$$\mathcal{S}(c) = \frac{1}{Z} \int e^{\sum_i c_i a_i} e^{W(a)} \, d\mu_S(a) \tag{1.5}$$

(called the generator of the Euclidean Green's functions) or equivalently, by

$$\mathcal{C}(c) = \log \frac{1}{Z} \int e^{\sum_i c_i a_i} e^{W(a)} \, d\mu_S(a) \tag{1.6}$$

(called the generator of the connected Green's functions) or equivalently (see Problem 1.14, below), by

$$\mathcal{G}(c) = \log \frac{1}{Z} \int e^{W(a+c)} \, d\mu_S(a) \tag{1.7}$$

(called the generator of the connected, freely amputated Green's functions).

PROBLEM 1.14. Prove that

$$\mathcal{C}(c) = -\frac{1}{2} \sum_{ij} c_i S_{ij} c_j + \mathcal{G}(-Sc)$$

where $(Sc)_i = \sum_j S_{ij} c_j$.

If S were really nice, the Grassmann Gaussian integrals of (1.5)–(1.7) would be very easy to define rigorously, even though the Grassmann algebra $\bigwedge \mathcal{V}$ is infinite-dimensional. The real problem is that S is not really nice. However, one can write S as the limit of really nice covariances. So the problem of making sense of the right hand side of, say, (1.7) may be expressed as the problem of controlling the limit of a sequence of well-defined quantities. The renormalization group is a tool that is used to control that limit. In the version of the renormalization group that I will use, the covariance S is written as a sum

$$S = \sum_{j=1}^{\infty} S^{(j)}$$

with each $S^{(j)}$ really nice. Let

$$S^{(\leq J)} = \sum_{j=1}^{J} S^{(j)}$$

and

$$\mathcal{G}_J(c) = \log \frac{1}{Z_j} \int e^{W(a+c)}\, d\mu_{S(\le J)}(a), \quad \text{where } Z_J = \int e^{W(a)}\, d\mu_{S(\le J)}(a).$$

(The normalization constant Z_J is chosen so that $\mathcal{G}_J(0) = 0$.) We have to control the limit of $\mathcal{G}_J(c)$ as $J \to \infty$. We have rigged the definitions so that it is very easy to exhibit the relationship between $\mathcal{G}_J(c)$ and $\mathcal{G}_{J+1}(c)$:

$$\begin{aligned}
\mathcal{G}_{J+1}(c) &= \log \frac{1}{Z_{J+1}} \int e^{W(b+c)}\, d\mu_{S(\le J+1)}(b)\\
&= \log \frac{1}{Z_{J+1}} \int e^{W(b+c)}\, d\mu_{S(\le J)+S(J+1)}(b)\\
&= \log \frac{1}{Z_{J+1}} \int \left[\int e^{W(a+b+c)}\, d\mu_{S(\le J)}(b)\right] d\mu_{S(J+1)}(a) \quad \text{by Proposition 1.21}\\
&= \log \frac{Z_1}{Z_{J+1}} \int e^{\mathcal{G}_J(c+a)}\, d\mu_{S(J+1)}(a).
\end{aligned}$$

PROBLEM 1.15. We have normalized $\mathcal{G}_{J+1}$ so that $\mathcal{G}_{J+1}(0) = 0$. So the ratio Z_j/Z_{J+1} in

$$\mathcal{G}_{J+1}(c) = \log \frac{Z_J}{Z_{J+1}} \int e^{\mathcal{G}_J(c+a)}\, d\mu_{S(J+1)}(a)$$

had better obey

$$\frac{Z_{J+1}}{Z_J} = \int e^{\mathcal{G}_J(a)}\, d\mu_{S(J+1)}(a).$$

Verify by direct computation that this is the case.

1.6. The Renormalization Group

By Problem 1.15,

$$\mathcal{G}_{J+1}(c) = \Omega_{S(J+1)}(\mathcal{G}_J)(c) \tag{1.8}$$

where the "renormalization group map" Ω is defined in

DEFINITION 1.22. Let $\mathcal{V}$ be a vector space with basis $\{c_i\}$. Choose a second copy $\mathcal{V}'$ of $\mathcal{V}$ and denote by a_i the basis element of $\mathcal{V}'$ corresponding to the element $c_i \in \mathcal{V}$. Let S be a skew symmetric bilinear form on $\mathcal{V}'$. The *renormalization group map* $\Omega_S\colon \bigwedge_{\mathbb{S}} \mathcal{V} \to \bigwedge_{\mathbb{S}} \mathcal{V}$ is defined by

$$\Omega_S(W)(c) = \log \frac{1}{Z_{W,S}} \int e^{W(c+a)}\, d\mu_S(a), \quad \text{where } Z_{W,S} = \int e^{W(a)}\, d\mu_S(a)$$

for all $W \in \bigwedge_{\mathbb{S}} \mathcal{V}$ for which $\int e^{W(a)}\, d\mu_S(a)$ is invertible in $\mathbb{S}$.

PROBLEM 1.16. Prove that

$$\begin{aligned}
\mathcal{G}_J &= \Omega_{S(J)} \circ \Omega_{S(J-1)} \circ \cdots \circ \Omega_{S(1)}(W)\\
&= \Omega_{S(1)} \circ \Omega_{S(2)} \circ \cdots \circ \Omega_{S(J)}(W).
\end{aligned}$$

For the rest of this section we restrict to $\mathbb{S} = \mathbb{C}$. Observe that we have normalized the renormalization group map so that $\Omega_S(W)(0) = 0$. Define the subspaces

$$\bigwedge^{(>0)} \mathcal{V} = \bigoplus_{n=1}^{\infty} \bigwedge^n \mathcal{V}$$

$$\bigwedge^{(e)} \mathcal{V} = \bigoplus_{\substack{n=2 \\ n \text{ even}}}^{\infty} \bigwedge^n \mathcal{V}$$

of $\bigwedge \mathcal{V}$ and the projection

$$\begin{aligned} P^{(>0)} \colon \bigwedge \mathcal{V} &\longrightarrow \bigwedge^{(>0)} \mathcal{V} \\ W(a) &\longmapsto W(a) - W(0). \end{aligned}$$

Then

LEMMA 1.23. (i) *If* $Z_{W,S} \neq 0$, *then* $Z_{P^{(>0)}W,S} \neq 0$ *and*

$$\Omega_S(W) = \Omega_S(P^{(>0)}W) \in \bigwedge^{(>0)} \mathcal{V}.$$

(ii) *If* $Z_{W,S} \neq 0$ *and* $W \in \bigwedge^{(e)} \mathcal{V}$, *then* $\Omega_S(W) \in \bigwedge^{(e)} \mathcal{V}$.

PROOF. (i) is an immediate consequence of

$$Z_{P^{(>0)}W,S} = e^{-W(0)} \int e^{W(a)}\, d\mu_S(a)$$

$$\int e^{(P^{(>0)}W)(c+a)}\, d\mu_S(a) = e^{-W(0)} \int e^{W(c+a)}\, d\mu_S(a).$$

(ii) Observe that $W \in \bigwedge \mathcal{V}$ is in $\bigwedge^{(e)} \mathcal{V}$ if and only if $W(0) = 0$ and $W(-a) = W(a)$. Suppose that $Z_{W,S} \neq 0$ and set $W' = \Omega_S(W)$. We already know that $W'(0) = 0$. Suppose also that $W(-a) = W(a)$. Then

$$W'(-c) = \ln \frac{\int e^{W(-c+a)}\, d\mu_S(a)}{\int e^{W(a)}\, d\mu_S(a)} = \ln \frac{\int e^{W(c-a)}\, d\mu_S(a)}{\int e^{W(a)}\, d\mu_S(a)}.$$

As $\int a_{\mathrm{I}}\, d\mu_S(a) = 0$ for all $|\mathrm{I}|$ odd we have

$$\int f(-a)\, d\mu_S(a) = \int f(a)\, d\mu_S(a)$$

for all $f(a) \in \bigwedge \mathcal{V}$ and

$$\ln \frac{\int e^{W(c-a)}\, d\mu_S(a)}{\int e^{W(a)}\, d\mu_S(a)} = \ln \frac{\int e^{W(c+a)}\, d\mu_S(a)}{\int e^{W(a)}\, d\mu_S(a)} = W'(c).$$

Hence $W'(-c) = W'(c)$. □

EXAMPLE 1.24. If $S = 0$, then $\Omega_{S=0}(W) = W$ for all $W \in \bigwedge^{(>0)} \mathcal{V}$. To see this, observe that $W(c+a) = W(c) + \widetilde{W}(c,a)$ with $\widetilde{W}(c,a)$ a polynomial in c and a with every term having degree at least one in a. When $S = 0$, $\int a_{\mathrm{I}}\, d\mu_S(a) = 0$ for all $|\mathrm{I}| \geq 1$ so that

$$\int W(c+a)^n\, d\mu_0(a) = \int [W(c) + \widetilde{W}(c,a)]^n\, d\mu_0(a) = W(c)^n$$

and

$$\int e^{W(c+a)}\, d\mu_0(a) = e^{W(c)}.$$

In particular

$$\int e^{W(a)}\,d\mu_0(a) = e^{W(0)} = 1.$$

Combining the last two equations

$$\ln \frac{\int e^{W(c+a)}\,d\mu_0(a)}{\int e^{W(a)}\,d\mu_0(a)} = \ln e^{W(c)} = W(c).$$

We conclude that

$$S = 0, W(0) = 0 \implies \Omega_S(W) = W.$$

EXAMPLE 1.25. Fix any $1 \le i, j \le D$ and $\lambda \in \mathbb{C}$ and let $W(a) = \lambda a_i a_j$. Then

$$\begin{aligned}\int e^{\lambda(c_i+a_i)(c_j+a_j)}\,d\mu_S(a) &= \int e^{\lambda c_i c_j} e^{\lambda a_i c_j} e^{\lambda c_i a_j} e^{\lambda a_i a_j}\,d\mu_S(a)\\ &= e^{\lambda c_i c_j}\int [1+\lambda a_i c_j][1+\lambda c_i a_j][1+\lambda a_i a_j]\,d\mu_S(a)\\ &= e^{\lambda c_i c_j}\int [1+\lambda a_i c_j+\lambda c_i a_j+\lambda a_i a_j+\lambda^2 a_i c_j c_i a_j]\,d\mu_S(a)\\ &= e^{\lambda c_i c_j}[1+\lambda S_{ij}+\lambda^2 c_j c_i S_{ij}].\end{aligned}$$

Hence, if $\lambda \ne -1/S_{ij}$, $Z_{W,S} = 1 + \lambda S_{ij}$ is nonzero and

$$\begin{aligned}\frac{\int e^{W(c+a)}\,d\mu_S(a)}{\int e^{W(a)}\,d\mu_S(a)} &= \frac{e^{\lambda c_i c_j}[1+\lambda S_{ij}+\lambda^2 c_j c_i S_{ij}]}{1+\lambda S_{ij}}\\ &= e^{\lambda c_i c_j}\left[1 + c_j c_i \frac{\lambda^2 S_{ij}}{1+\lambda S_{ij}}\right]\\ &= e^{\lambda c_i c_j - \lambda^2 S_{ij} c_i c_j/(1+\lambda S_{ij})}\\ &= e^{\lambda c_i c_j/(1+\lambda S_{ij})}.\end{aligned}$$

We conclude that

$$W(a) = \lambda a_i a_j, \qquad \lambda \ne -\frac{1}{S_{ij}} \implies \Omega_S(W)(c) = \frac{\lambda}{1+\lambda S_{ij}} c_i c_j.$$

In particular $Z_{0,S} = 1$ and

$$\Omega_S(0) = 0$$

for all S.

DEFINITION 1.26. Let $\mathcal{E}$ be any finite-dimensional vector space and let $f\colon \mathcal{E} \to \mathbb{C}$ and $F\colon \mathcal{E} \to \mathcal{E}$. For each ordered basis $B = (e_1, \ldots, e_D)$ of $\mathcal{E}$ define $\tilde{f}_B\colon \mathbb{C}^D \to \mathbb{C}$ and $\widetilde{F}_{B;i}\colon \mathbb{C}^D \to \mathbb{C}$, $1 \le i \le D$ by

$$\begin{aligned}f(\lambda_1 e_1 + \cdots + \lambda_D e_D) &= \tilde{f}_B(\lambda_1, \ldots, \lambda_D)\\ F(\lambda_1 e_1 + \cdots + \lambda_D e_D) &= \widetilde{F}_{B;1}(\lambda_1, \ldots, \lambda_D)e_1 + \cdots + \widetilde{F}_{B;D}(\lambda_1, \ldots, \lambda_D)e_D.\end{aligned}$$

We say that f is polynomial (rational) if there exists a basis B for which $\tilde{f}_B(\lambda_1, \ldots, \lambda_D)$ is a polynomial (ratio of two polynomials). We say that F is polynomial (rational) if there exists a basis B for which $\widetilde{F}_{B;i}(\lambda_1, \ldots, \lambda_D)$ is a polynomial (ratio of two polynomials) for all $1 \le i \le D$.

REMARK 1.27. If f is polynomial (rational) then $\tilde{f}_B(\lambda_1, \ldots, \lambda_D)$ is a polynomial (ratio of two polynomials) for all bases B. If F is polynomial (rational) then $\widetilde{F}_{B;i}(\lambda_1, \ldots, \lambda_D)$ is a polynomial (ratio of two polynomials) for all $1 \le i \le D$ and all bases B.

PROPOSITION 1.28. *Let $\mathcal{V}$ be a finite-dimensional vector space and fix any skew symmetric bilinear form S on $\mathcal{V}$. Then*

(i)

$$\begin{aligned} Z_{W,S} \colon \textstyle\bigwedge^{(>0)} \mathcal{V} &\longrightarrow \mathbb{C} \\ W(a) &\longmapsto \int e^{W(a)}\, d\mu_S(a) \end{aligned}$$

is polynomial.

(ii)

$$\begin{aligned} \Omega_S(W) \colon \textstyle\bigwedge^{(>0)} \mathcal{V} &\longrightarrow \textstyle\bigwedge^{(>0)} \mathcal{V} \\ W(a) &\longmapsto \ln \frac{\int e^{W(a+c)}\, d\mu_S(c)}{\int e^{W(c)}\, d\mu_S(c)} \end{aligned}$$

is rational.

PROOF. (i) Let D be the dimension of $\mathcal{V}$. If $W(0) = 0$, then $e^{W(a)} = \sum_{n=0}^{D} W(a)^D/n!$ is polynomial in W. Hence, so is $Z_{W,S}$.

(ii) As in part (i), $\int e^{W(a+c)}\, d\mu_S(c)$ and $\int e^{W(c)}\, d\mu_S(c)$ are both polynomial. By Example 1.25, $\int e^{W(c)}\, d\mu_S(c)$ is not identically zero. Hence

$$\widetilde{W}(a) = \frac{\int e^{W(a+c)}\, d\mu_S(c)}{\int e^{W(c)}\, d\mu_S(c)}$$

is rational and obeys $\widetilde{W}(0) = 1$. Hence

$$\ln \widetilde{W} = \sum_{n=1}^{D} \frac{(-1)^{(n-1)}}{n} (\widetilde{W} - 1)^n$$

is polynomial in $\widetilde{W}$ and rational in W. □

THEOREM 1.29. *Let $\mathcal{V}$ be a finite-dimensional vector space.*

(i) *If S and T are any two skew symmetric bilinear forms on $\mathcal{V}$, then $\Omega_S \circ \Omega_T$ is defined as a rational function and*

$$\Omega_S \circ \Omega_T = \Omega_{S+T}.$$

(ii) *$\{\Omega_S(\cdot) \mid S$ a skew symmetric bilinear form on $\mathcal{V}\}$ is an abelian group under composition and is isomorphic to $\mathbb{R}^{D(D-1)/2}$, where D is the dimension of $\mathcal{V}$.*

PROOF. (i) To verify that $\Omega_S \circ \Omega_T$ is defined as a rational function, we just need to check that the range of Ω_T is not contained in the zero set of $Z_{W,S}$. This is the case because $\Omega_T(0) = 0$, so that 0 is in the range of Ω_T, and $Z_{0,S} = 1$.

As $Z_{W,T}$, $Z_{\Omega_T(W),S}$ and $Z_{W,S+T}$ are all rational functions of W that are not identically zero

$$\{W \in \textstyle\bigwedge \mathcal{V} \mid Z_{W,T} \neq 0, Z_{\Omega_T(W),S} \neq 0, Z_{W,S+T} \neq 0\}$$

is an open dense subset of $\bigwedge \mathcal{V}$. On this subset

$$\begin{aligned}\Omega_{S+T}(W) &= \ln \frac{\int e^{W(c+a)}\, d\mu_{S+T}(a)}{\int e^{W(a)}\, d\mu_{S+T}(a)} \\ &= \ln \frac{\int \left[\int e^{W(a+b+c)}\, d\mu_T(b)\right] d\mu_S(a)}{\int \left[\int e^{W(a+b)}\, d\mu_T(b)\right] d\mu_S(a)} \\ &= \ln \frac{\int e^{\Omega_T(W)(a+c)} \int e^{W(b)}\, d\mu_T(b)\, d\mu_S(a)}{\int e^{\Omega_T(W)(a)} \int e^{W(b)}\, d\mu_T(b)\, d\mu_S(a)} \\ &= \ln \frac{\int e^{\Omega_T(W)(a+c)}\, d\mu_S(a)}{\int e^{\Omega_T(W)(a)}\, d\mu_S(a)} \\ &= \Omega_S\big(\Omega_T(W)\big).\end{aligned}$$

(ii) The additive group of $D \times D$ skew symmetric matrices is isomorphic to $\mathbb{R}^{D(D-1)/2}$. By part (i), $S \mapsto \Omega_S(\cdot)$ is a homomorphism from the additive group of $D \times D$ skew symmetric matrices onto $\{\Omega_S(\cdot) \mid S \text{ a } D \times D \text{ skew symmetric matrix}\}$. To verify that it is an isomorphism, we just need to verify that the map is one-to-one. Suppose that

$$\Omega_S(W) = \Omega_T(W)$$

for all W with $Z_{W,S} \neq 0$ and $Z_{W,T} \neq 0$. Then, by Example 1.25, for each $1 \leq i, j \leq D$,

$$\frac{\lambda}{1 + \lambda S_{ij}} = \frac{\lambda}{1 + \lambda T_{ij}}$$

for all $\lambda \neq -1/S_{ij}, -1/T_{ij}$. But this implies that $S_{ij} = T_{ij}$.

□

1.7. Wick Ordering

Let $\mathcal{V}$ be a D-dimensional vector space over $\mathbb{C}$ and let $\mathbb{S}$ be a superalgebra. Let S be a $D \times D$ skew symmetric matrix. We now introduce a new basis for $\bigwedge_{\mathbb{S}} \mathcal{V}$ that is adapted for use with the Grassmann Gaussian integral with covariance S. To this point, we have always selected some basis $\{a_1, \ldots, a_D\}$ for $\mathcal{V}$ and used $\{a_{i_1} \cdots a_{i_n} \mid n \geq 0, 1 \leq i_1 < \cdots < i_n \leq D\}$ as a basis for $\bigwedge_{\mathbb{S}} \mathcal{V}$. The new basis will be denoted

$$\{{:}a_{i_1} \cdots a_{i_n}{:} \mid n \geq 0, 1 \leq i_1 < \cdots < i_n \leq D\}.$$

The basis element ${:}a_{i_1} \cdots a_{i_n}{:}$ will be called the Wick ordered product of $a_{i_1} \cdots a_{i_n}$.

Let $\mathcal{V}'$ be a second copy of $\mathcal{V}$ with basis $\{b_1, \ldots, b_D\}$. Then ${:}a_{i_1} \cdots a_{i_n}{:}$ is determined by applying $\partial/\partial b_{i_1} \cdots \partial/\partial b_{i_n}$ to both sides of

$${:}e^{\sum b_i a_i}{:} = e^{(\sum b_i S_{ij} b_j)/2} e^{\sum b_i a_i}$$

and setting all of b_i's to zero. This determines Wick ordering as a linear map on $\bigwedge_{\mathbb{S}} \mathcal{V}$. Clearly, the map depends on S, even though we have not included any S in the notation. It is easy to see that ${:}1{:} = 1$ and that ${:}a_{i_1} \cdots a_{i_n}{:}$ is

- a polynomial in $a_{i_1}, \ldots, a_{i_n}$ of degree n with degree n term precisely $a_{i_1} \cdots a_{i_n}$
- an even, resp. odd, element of $\bigwedge_{\mathbb{S}} \mathcal{V}$ when n is even, resp. odd.
- antisymmetric under permutations of $i_1, \ldots, i_n$ (because $\partial/\partial b_{i_1} \cdots \partial/\partial b_{i_n}$ is antisymmetric under permutations of $i_1, \ldots, i_n$).

So the linear map $f(a) \mapsto {:}f(a){:}$ is bijective. The easiest way to express $a_{i_1} \cdots a_{i_n}$ in terms of the basis of Wick ordered monomials is to apply $\partial/\partial b_{i_n} \cdots \partial/\partial b_{i_n}$ to both sides of

$$e^{\sum b_i a_i} = e^{-(\sum b_i S_{ij} b_j)/2} {:}e^{\sum b_i a_i}{:}$$

and set all of the b_i's to zero.

EXAMPLE 1.30.

$$\begin{aligned} {:}a_i{:} &= a_i & a_i &= {:}a_i{:} \\ {:}a_i a_j{:} &= a_i a_j - S_{ij} & a_i a_j &= {:}a_i a_j{:} + S_{ij} \end{aligned}$$

$$\begin{aligned} {:}a_i a_j a_k{:} &= a_i a_j a_k - S_{ij} a_k - S_{ki} a_j - S_{jk} a_i \\ a_i a_j a_k &= {:}a_i a_j a_k{:} + S_{ij}{:}a_k{:} + S_{ki}{:}a_j{:} + S_{jk}{:}a_i{:}. \end{aligned}$$

By the antisymmetry property mentioned above, ${:}a_i \cdots a_{i_n}{:}$ vanishes if any two of the indices i_j are the same. However ${:}a_{i_1} \cdots a_{i_n}{:}\ {:}a_{j_1} \cdots a_{j_m}{:}$ need not vanish if one of the i_k's is the same as one of the j_ℓ's. For example

$${:}a_i a_j{:}\ {:}a_i a_j{:} = (a_i a_j - S_{ij})(a_i a_j - S_{ij}) = -2S_{ij} a_i a_j + S_{ij}^2.$$

PROBLEM 1.17. Prove that

$$\begin{aligned} {:}f(a){:} &= \int f(a+b)\, d\mu_{-S}(b) \\ f(a) &= \int {:}f{:}(a+b)\, d\mu_S(b). \end{aligned}$$

PROBLEM 1.18. Prove that

$$\frac{\partial}{\partial a_\ell} {:}f(a){:} = {:}\frac{\partial}{\partial a_\ell} f(a){:}.$$

PROBLEM 1.19. Prove that

$$\int {:}g(a) a_i{:} f(a)\, d\mu_S(a) = \sum_{\ell=1}^{D} S_{i\ell} \int {:}g(a){:} \frac{\partial}{\partial a_\ell} f(a)\, d\mu_S(a).$$

PROBLEM 1.20. Prove that

$${:}f{:}(a+b) = {:}f(a+b){:}_a = {:}f(a+b){:}_b.$$

Here ${:}\cdot{:}_a$ means Wick ordering of the a_i's and ${:}\cdot{:}_b$ means Wick ordering of the b_i's. Precisely, if $\{a_i\}$, $\{b_i\}$, $\{A_i\}$, $\{B_i\}$ are bases of for four vector spaces, all of the same dimension,

$$\begin{aligned} {:}e^{\sum_i A_i a_i + \sum_i B_i b_i}{:}_a &= e^{\sum_i B_i b_i} {:}e^{\sum_i A_i a_i}{:}_a = e^{\sum_i A_i a_i + \sum_i B_i b_i} e^{(\sum_{ij} A_i S_{ij} A_j)/2} \\ {:}e^{\sum_i A_i a_i + \sum_i B_i b_i}{:}_b &= e^{\sum_i A_i a_i} {:}e^{\sum_i B_i b_i}{:}_b = e^{\sum_i A_i a_i + \sum_i B_i b_i} e^{(\sum_{ij} B_i S_{ij} B_j)/2}. \end{aligned}$$

PROBLEM 1.21. Prove that

$${:}f{:}_{S+T}(a+b) = {:}f(a+b){:}_{\substack{a,S \\ b,T}}.$$

Here S and T are skew symmetric matrices, ${:}\cdot{:}_{S+T}$ means Wick ordering with respect to $S+T$ and ${:}\cdot{:}_{\substack{a,S \\ b,T}}$ means Wick ordering of the a_i's with respect to S and of the b_i's with respect to T.

PROPOSITION 1.31 (Orthogonality of Wick Monomials). (a) *If* $m > 0$

$$\int :a_{i_1} \cdots a_{i_m}: d\mu_S(a) = 0$$

(b) *If* $m \neq n$

$$\int :a_{i_1} \cdots a_{i_m}::a_{j_i} \cdots a_{j_n}: d\mu_S(a) = 0$$

(c)

$$\int :a_{i_1} \cdots a_{i_m}::a_{j_m} \cdots a_{j_1}: d\mu_S(a) = \det[S_{i_k j_\ell}]_{l \le k,\ell \le m}.$$

Note the order of the indices in $:a_{j_m} \cdots a_{j_1}:$.

PROOF. (a), (b) Let $\mathcal{V}''$ be a third copy of $\mathcal{V}$ with basis $\{c_1, \ldots, c_D\}$. Then

$$\begin{aligned}\int :e^{\sum b_i a_i}::e^{\sum c_i a_i}: d\mu_S(a) &= \int e^{1/2 \sum b_i S_{ij} b_j} e^{\sum b_i a_i} e^{(\sum c_i S_{ij} c_j)/2} e^{\sum c_i a_i}\, d\mu_S(a) \\ &= e^{(\sum b_i S_{ij} b_j)/2} e^{(\sum c_i S_{ij} c_j)/2} \int e^{\sum (b_i + c_i) a_i}\, d\mu_S(a) \\ &= e^{(\sum b_i S_{ij} b_j)/2} e^{(\sum c_i S_{ij} c_j)/2} e^{-[\sum (b_i + c_i) S_{ij} (b_j + c_j)]/2} \\ &= e^{-\sum b_i S_{ij} c_j}.\end{aligned}$$

Now apply $\partial/\partial b_{i_1} \ldots \partial/\partial b_{i_m}$ and set all of the b_i's to zero

$$\begin{aligned}&\frac{\partial}{\partial b_{i_1}} \cdots \frac{\partial}{\partial b_{i_m}} e^{-\sum b_i S_{ij} c_j}\Big|_{b_i=0} \\ &\qquad = e^{-\sum b_i S_{ij} c_j} \Big(-\sum_{j'_1} S_{i_1 j'_1} c_{j'_1}\Big) \cdots \Big(-\sum_{j'_m} S_{i_m j'_m} c_{j'_m}\Big)\Big|_{b_i=0} \\ &\qquad = (-1)^m \Big(\sum_{j'_1} S_{i_1 j'_1} c_{j'_1}\Big) \cdots \Big(\sum_{j'_m} S_{i_m j'_m} c_{j'_m}\Big).\end{aligned}$$

Now apply $\partial/\partial c_{j_1} \cdots \partial/\partial c_{j_n}$ and set all of the c_i's to zero. To get a nonzero answer, it is necessary that $m = n$.

(c) We shall prove that

$$\int :a_{i_m} \cdots a_{i_1}::a_{j_1} \ldots a_{j_m}: d\mu_S(a) = \det[S_{i_k j_\ell}]_{1 \le k,\ell \le m}.$$

This is equivalent to the claimed result. By Problems 1.19 and 1.18,

$$\begin{aligned}&\int :a_{i_m} \cdots a_{i_1}::a_{j_1} \cdots a_{j_m}: d\mu_S(a) \\ &\qquad = \sum_{\ell=1}^{m} (-1)^{\ell+1} S_{i_1 j_\ell} \int :a_{i_m} \cdots a_{i_2}::\prod_{\substack{k=1\\k\neq\ell}}^{m} a_{j_k}: d\mu_S(a).\end{aligned}$$

The proof will be by induction on m. If $m = 1$, we have

$$\int :a_{i_1}::a_{j_1}: d\mu_S(a) = S_{i_1 j_1} \int 1\, d\mu_S(a) = S_{i_1 j_1}$$

as desired. In general, by the inductive hypothesis,

$$\begin{aligned}\int :a_{i_m}\cdots a_{i_1}::a_{j_1}\cdots a_{j_m}:d\mu_S(a) &= \sum_{\ell=1}^{m}(-1)^{\ell+1}S_{i_1 j_\ell}\det[S_{i_p j_k}]_{\substack{1\le p,k\le m\\ p\ne 1,k\ne \ell}}\\ &= \det[S_{i_p j_k}]_{1\le k,p\le m}.\end{aligned}$$

□

PROBLEM 1.22. Prove that

$$\int :f(a):d\mu_S(a) = f(0)$$

PROBLEM 1.23. Prove that

$$\int \prod_{i=1}^{n} :\prod_{\mu=1}^{e_i} a_{\ell_{i,\mu}}:d\mu_S(\psi) = \mathrm{Pf}\big(T_{(i,\mu),(i',\mu')}\big)$$

where

$$T_{(i,\mu),(i',\mu')} = \begin{cases}0, & \text{if } i=i';\\ S_{\ell_{i,\mu},\ell_{i',\mu'}}, & \text{if } i\ne i'.\end{cases}$$

Here T is a skew symmetric matrix with $\sum_{i=1}^{n} e_i$ rows and columns, numbered, in order $(1,1),\ldots,(1,e_1),(2,1),\ldots,(2,e_2),\ldots,(n,e_n)$. The product in the integrand is also in this order. Hint: Use Problems 1.19 and 1.18 and Proposition 1.18.

1.8. Bounds on Grassmann Gaussian Integrals

We now prove some bounds on Grassmann Gaussian integrals. While it is not really necessary to do so, I will make some simplifying assumptions that are satisfied in applications to quantum field theories. I will assume that the vector space $\mathcal{V}$ generating the Grassmann algebra has basis $\{\psi(\ell,\kappa)\mid \ell\in X, \kappa\in\{0,1\}\}$, where X is some finite set. Here, $\psi(\ell,0)$ plays the role of $\psi_{x,\sigma}$ of Section 1.5 and $\psi(\ell,1)$ plays the role of $\bar\psi_{x,\sigma}$ of Section 1.5. I will also assume that, as in (1.4), the covariance only couples $\kappa=0$ generators to $\kappa=1$ generators. In other words, we let A be a function on $X\times X$ and consider the Grassmann Gaussian integral $\int\cdot\, d\mu_A(\psi)$ on $\bigwedge\mathcal{V}$ with

$$\int \psi(\ell,\kappa)\psi(\ell',\kappa')\,d\mu_A(\psi) = \begin{cases}0, & \text{if } \kappa=\kappa'=0;\\ A(\ell,\ell'), & \text{if } \kappa=0,\ \kappa'=1;\\ -A(\ell,\ell'), & \text{if } \kappa=1,\ \kappa'=0;\\ 0, & \text{if } \kappa=\kappa'=1.\end{cases}$$

We start off with the simple bound.

PROPOSITION 1.32. *Assume that there is Hilbert space $\mathcal{H}$ and vectors f_ℓ, g_ℓ, $\ell\in X$ in $\mathcal{H}$ such that*

$$A(\ell,\ell') = \langle f_\ell, g_{\ell'}\rangle_{\mathcal{H}} \quad \textit{for all } \ell,\ell'\in X.$$

Then

$$\left|\int \prod_{i=1}^{n}\psi(\ell_i,\kappa_i)\,d\mu_A(\psi)\right| \le \prod_{\substack{i=1\\ \kappa_i=0}}^{n}\|f_{\ell_i}\|_{\mathcal{H}}\prod_{\substack{i=1\\ \kappa_i=1}}^{n}\|g_{\ell_i}\|_{\mathcal{H}}.$$

PROOF. Define

$$F = \{1 \le i \le n \mid \kappa_i = 0\}$$
$$\overline{F} = \{1 \le i \le n \mid \kappa_i = 1\}.$$

By Problem 1.13, if the integral does not vanish, the ordinality of F and $\overline{F}$ coincide and there is a sign $\pm$ such that

$$\int \prod_{i=1}^{n} \psi(\ell_i, \kappa_i)\, d\mu_A(\psi) = \pm \det[A_{\ell_i,\ell_j}]_{\substack{i\in F\\ j\in\overline{F}}}.$$

The proposition is thus an immediate consequence of Gram's inequality. For the convenience of the reader, we include a proof of this classical inequality below. □

LEMMA 1.33 (Gram's inequality). *Let $\mathcal{H}$ be a Hilbert space and $u_1, \ldots, u_n$, $v_1, \ldots, v_n \in \mathcal{H}$. Then*

$$|\det[\langle u_i, v_j\rangle]_{1\le i,j\le n}| \le \prod_{i=1}^{n} \|u_i\|\,\|v_i\|.$$

Here, $\langle\cdot,\cdot\rangle$ and $\|\cdot\|$ are the inner product and norm in $\mathcal{H}$, respectively.

PROOF. We start with three reductions. First, we may assume that the $u_1, \ldots, u_n$ are linearly independent. Otherwise the determinant vanishes, because its rows are not independent, and the inequality is trivially satisfied. Second, we may also assume that each v_j is in the span of the u_i's, because, if P is the orthogonal projection onto that span, $\det[\langle u_i, v_j\rangle]_{1\le i,j\le n} = \det[\langle u_i, Pv_j\rangle]_{1\le i,j\le n}$ while $\prod_{i=1}^{n} \|Pu_i\|\,\|v_i\| \le \prod_{i=1}^{n} \|u_i\|\,\|v_i\|$. Third, we may assume that $v_1, \ldots, v_n$ are linearly independent. Otherwise the determinant vanishes, because its columns are not independent. Denote by U the span of $u_1, \ldots, u_n$. We have just shown that we may assume that $u_1, \ldots, u_n$ and $v_1, \ldots, v_n$ are two bases for U.

Let α_i be the projection of u_i on the orthogonal complement of the subspace spanned by $u_1, \ldots, u_{i-1}$. Then $\alpha_i = u_i = +\sum_{j=1}^{i-1} \tilde{L}_{ij} u_j$ for some complex numbers $\tilde{L}_{ij}$ and α_i is orthogonal to $u_1, \ldots, u_{i-1}$ and hence to $\alpha_1, \ldots, \alpha_{i-1}$. Set

$$L_{ij} = \begin{cases} \|\alpha_i\|^{-1}, & \text{if } i = j; \\ 0, & \text{if } i < j; \\ \|\alpha_i\|^{-1}\tilde{L}_{ij}, & \text{if } i > j. \end{cases}$$

Then L is a lower triangular matrix with diagonal entries

$$L_{ii} = \|\alpha_i\|^{-1}.$$

such that the linear combinations

$$u_i' = \sum_{j=1}^{i} L_{ij} u_j, \quad i = 1, \ldots, n$$

are orthonormal. This is just the Gram-Schmidt orthogonalization algorithm. Similarly, let β_i be the projection of v_i on the orthogonal complement of the subspace spanned by $v_1, \ldots, v_{i-1}$. By Gram-Schmidt, there is a lower triangular matrix M with diagonal entries

$$M_{ii} = \|\beta_i\|^{-1}$$

such that the linear combinations

$$v_i' = \sum_{j=1}^{i} M_{ij} v_j, \quad i = 1, \dots, n$$

are orthonormal. Since the v_i''s are orthonormal and have the same span as the v_i's, they form an orthonormal basis for U. As a result, $u_i' = \sum_j \langle u_i', v_j' \rangle v_j'$ so that

$$\sum_j \langle u_i', v_j' \rangle \langle v_j', u_k' \rangle = \langle u_i', u_k' \rangle = \delta_{i,k}$$

and the matrix $[\langle u_i', v_j' \rangle]$ is orthogonal and consequently has determinant one. As

$$L[\langle u', v_j \rangle] M^\dagger = [\langle v_i', v_j' \rangle]$$

we have

$$\det[\langle u_i, v_j \rangle] = \det{}^{-1} L \, \det{}^{-1} M = \prod_{i=1}^{n} \|\alpha_i\| \, \|\beta_i\| \le \prod_{i=1}^{n} \|u_i\| \, \|v_i\|$$

since

$$\begin{aligned} \|u_j\|^2 &= \|\alpha_j + (u_j - \alpha_j)\|^2 = \|\alpha_j\|^2 + \|u_j - \alpha_j\|^2 \ge \|\alpha_j\|^2 \\ \|v_j\|^2 &= \|\beta_j + (v_j - \beta_j)\|^2 = \|\beta_j\|^2 + \|v_j - \beta_j\|^2 \ge \|\beta_j\|^2. \end{aligned}$$

Here, we have used that α_j and $u_j - \alpha_j$ are orthogonal, because α_j is the orthogonal projection of u_j on some subspace. Similarly, β_j and $v_j - \beta_j$ are orthogonal. □

PROBLEM 1.24. Let a_{ij}, $1 \le i, j \le n$ be complex numbers. Prove Hadamard's inequality

$$|\det[a_{ij}]| \le \prod_{i=1}^{n} \Big(\sum_{j=1}^{n} |a_{ij}|^2 \Big)^{1/2}$$

from Gram's inequality. (Hint: Express $a_{ij} = \langle a_i, e_j \rangle$ where $e_j = (0, \dots, 1, \dots, 0)$, $j = 1, \dots, n$ is the standard basis for $\mathbb{C}^n$.)

PROBLEM 1.25. Let $\mathcal{V}$ be a vector space with basis $\{a_1, \dots, a_D\}$. Let $S_{\ell,\ell'}$ be a skew symmetric $D \times D$ matrix with

$$S(\ell, \ell') = \langle f_\ell, g_{\ell'} \rangle_{\mathcal{H}} \quad \text{for all } 1 \le \ell, \ell' \le D$$

for some Hilbert space $\mathcal{H}$ and vectors $f_\ell, g_{\ell'} \in \mathcal{H}$. Set $F_\ell = \sqrt{\|f_\ell\|_{\mathcal{H}} \|g_\ell\|_{\mathcal{H}}}$. Prove that

$$\left| \int \prod_{\ell=1}^{n} a_{i_\ell} \, d\mu_S(a) \right| \le \prod_{\ell=1}^{n} F_{i_\ell}.$$

We shall need a similar bound involving a Wick monomial. Let $:\cdot:$ denote Wick ordering with respect to $d\mu_A$.

PROPOSITION 1.34. *Assume that there is a Hilbert space $\mathcal{H}$ and vectors f_ℓ, g_ℓ, $\ell \in X$ in $\mathcal{H}$ such that*

$$A(\ell, \ell') = \langle f_\ell, g_{\ell'} \rangle_{\mathcal{H}} \quad \textit{for all } \ell, \ell' \in X.$$

Then

$$\left|\int \prod_{i=1}^{m} \psi(\ell_i,\kappa_i) :\prod_{j=1}^{n} \psi(\ell'_j,\kappa'_j): d\mu_A(\psi)\right| \le 2^n \prod_{\substack{i=1\\ \kappa_i=0}}^{m} \|f_{\ell_1}\|_{\mathcal{H}} \prod_{\substack{i=1\\ \kappa_i=1}}^{m} \|g_{\ell_i}\|_{\mathcal{H}} \prod_{\substack{j=1\\ \kappa'_j=0}}^{n} \|f_{\ell'_j}\|_{\mathcal{H}} \prod_{\substack{j=1\\ \kappa'_j=1}}^{n} \|g_{\ell'_j}\|_{\mathcal{H}}.$$

PROOF. By Problem 1.17

$$\begin{aligned}\int \prod_{i=1}^{m} \psi(\ell_i,\kappa_i) :\prod_{j=1}^{n} \psi(\ell'_j,\kappa'_j): d\mu_A(\psi) &= \int \prod_{i=1}^{m} \psi(\ell_i,\kappa_i) \prod_{j=1}^{n} [\psi(\ell'_j,\kappa'_j) + \phi(\ell'_j\kappa'_j)]\, d\mu_A(\psi)\, d\mu_{-A}(\phi) \\ &= \sum_{J\subset\{1,\dots,n\}} \pm \int \prod_{i=1}^{m}\prod_{j\in J} \psi(\ell'_j,\kappa'_j)\, d\mu_A(\psi) \int \prod_{j\notin J} \phi(\ell'_j,\kappa'_j)\, d\mu_{-A}(\phi).\end{aligned}$$

There are 2^n terms in this sum. For each term, the first factor is, by Proposition 1.32, no larger than

$$\prod_{\substack{i=1\\ \kappa_i=0}}^{m} \|f_{\ell_i}\|_{\mathcal{H}} \prod_{\substack{i=1\\ \kappa_i=1}}^{m} \|g_{\ell_i}\|_{\mathcal{H}} \prod_{\substack{j\in J\\ \kappa'_j=0}} \|f_{\ell'_i}\|_{\mathcal{H}} \prod_{\substack{j\in J\\ \kappa'_j=1}} \|g_{\ell'_j}\|_{\mathcal{H}}$$

and the second factor is, again by Proposition 1.32 (since $-A(\ell,\ell') = \langle f_\ell, -g_{\ell'}\rangle_{\mathcal{H}}$), no larger than

$$\prod_{\substack{j\notin J\\ \kappa'_j=0}} \|f_{\ell'_j}\|_{\mathcal{H}} \prod_{\substack{j\notin J\\ \kappa'_j=1}} \|g_{\ell'_j}\|_{\mathcal{H}}.$$

□

PROBLEM 1.26. Prove, under the hypotheses of Proposition 1.34, that

$$\left|\int \prod_{i=1}^{n} :\prod_{\mu=1}^{e_i} \psi(\ell_{i,\mu},\kappa_{i,\mu}): d\mu_A(\psi)\right| \le \prod_{\substack{1\le i<n\\ 1\le\mu\le e_i\\ \kappa_{i,\mu}=0}} \sqrt{2}\|f_{\ell_{i,\mu}}\|_{\mathcal{H}} \prod_{\substack{1\le i\le n\\ 1\le\mu\le e_i\\ \kappa_{i,\mu}=1}} \sqrt{2}\|g_{\ell_{i,\mu}}\|_{\mathcal{H}}.$$

Finally, we specialize to almost the full structure of (1.4). We only replace $\delta_{\sigma,\sigma'}/\big(ik_0 - e(\mathbf{k})\big)$ by a general matrix valued function of the (momentum) k. This is the typical structure in quantum field theories. Let $\mathfrak{G}$ be a finite set. Let $E_{\sigma,\sigma'}(k) \in L^1\big(\mathbb{R}^{d+1}, dk/(2\pi)^{d+1}\big)$, for each $\sigma,\sigma' \in \mathfrak{G}$, and set

$$C_{\sigma,\sigma'}(x,y) = \int \frac{d^{d+1}k}{(2\pi)^{d+1}} e^{ik\cdot(y-x)} E_{\sigma,\sigma'}(k).$$

Let $\int \cdot\, d\mu_C(\psi)$ be the Grassmann Gaussian integral determined by

$$\int \psi_\sigma(x,\kappa)\psi_{\sigma'}(y,\kappa')\, d\mu_C(\psi) = \begin{cases} 0, & \text{if } \kappa=\kappa'=0; \\ C_{\sigma,\sigma'}(x,y), & \text{if } \kappa=0,\ \kappa'=1; \\ -C_{\sigma,\sigma'}(y,x), & \text{if } \kappa=1,\ \kappa'=0; \\ 0, & \text{if } \kappa=\kappa'=1, \end{cases}$$

for all $x, y \in \mathbb{R}^{d+1}$ and $\sigma, \sigma' \in \mathfrak{G}$.

COROLLARY 1.35.

$$\sup_{\substack{x_i,\sigma_i,\kappa_i\\ x'_j,\sigma'_j,\kappa'_j}} \left| \int \prod_{i=1}^{m} \psi_{\sigma_i}(x_i,\kappa_i) :\prod_{j=1}^{n} \psi_{\sigma'_j}(x'_j,\kappa'_j): d\mu_C(\psi) \right| \le 2^n \left(\int \|E(k)\| \frac{dk}{(2\pi)^{d+1}} \right)^{(m+n)/2}$$

$$\sup_{\substack{x_{i,\mu}\sigma_{i,\mu}\\ \kappa_{i,\mu}}} \left| \int \prod_{i=1}^{n} :\psi_{\sigma_{i,1}}(x_{i,1},\kappa_{i,1}) \cdots \psi_{\sigma_{i,e_i}}(x_{i,e_i},\kappa_{i,e_i}): d\mu_C(\psi) \right| \le \left(2 \int \|E(k)\| \frac{dk}{(2\pi)^{d+1}} \right)^{\sum_i e_i/2}.$$

Here $\|E(k)\|$ *denotes the norm of the matrix* $\big(E_{\sigma,\sigma'}(k)\big)_{\sigma,\sigma'\in\mathfrak{G}}$ *as an operator on* $\ell^2(\mathbb{C}^{|\mathfrak{G}|})$.

PROOF.

$$X = \{(i,\mu) \mid 1 \le i \le n, 1 \le \mu \le e_i\}$$
$$A\big((i,\mu),(i',\mu')\big) = C_{\sigma_{i,\mu},\sigma_{i',\mu'}}(x_{i,\mu}, x_{i',\mu'}).$$

Let $\Psi\big((i,\mu),\kappa\big)$, $(i,\mu) \in X$, $\kappa \in \{0,1\}$ be generators of a Grassmann algebra and let $d\mu_A(\Psi)$ be a Grassmann Gaussian measure on that algebra with covariance A. This construction has been arranged so that

$$\int \psi_{\sigma_{i,\mu}}(x_{i,\mu},\kappa_{i,\mu}) \psi_{\sigma_{i',\mu'}}(x_{i',\mu'},\kappa_{i',\mu'})\, d\mu_C(\psi) = \int \Psi\big((i,\mu),\kappa_{i,\mu}\big) \Psi\big((i',\mu'),\kappa_{i',\mu'}\big)\, d\mu_A(\Psi)$$

and consequently

$$\int \prod_{i=1}^{n} :\psi_{\sigma_{i,1}}(x_{i,1},\kappa_{i,1}) \ldots \psi_{\sigma_{i,e_i}},(x_{i,e_i},\kappa_{i,e_i}): d\mu_C(\psi) = \int \prod_{i=1}^{n} :\Psi\big((i,1),\kappa_{i,1}\big) \ldots \Psi\big((i,e_i),\kappa_{i,e_i}\big): d\mu_A(\Psi).$$

Let $\mathcal{H} = L^2(\mathbb{R}^{d+1}, dk/(2\pi)^{d+1}) \otimes \mathbb{C}^{|\mathfrak{G}|}$ and

$$f_{i,\mu}(k,\sigma) = e^{ik\cdot x_{i,\mu}} \sqrt{\|E(k)\|}\, \delta_{\sigma,\sigma_{i,\mu}}, \qquad g_{i,\mu}(k,\sigma) = e^{ik\cdot x_{i,\mu}} \frac{E_{\sigma,\sigma_{i,\mu}}(k)}{\sqrt{\|E(k)\|}}.$$

If $\|E(k)\| = 0$, set $g_{i,\mu}(k,\sigma) = 0$. Then

$$A\big((i,\mu),(i',\mu')\big) = \langle f_{i,\mu}, g_{i',\mu'} \rangle_{\mathcal{H}}$$

and, since $\sum_{\sigma\in\mathfrak{G}} |E_{\sigma,\sigma_{i,\mu}}(k)|^2 \le \|E(k)\|^2$,

$$\|f_{i,\mu}\|_{\mathcal{H}}, \|g_{i,\mu}\|_{\mathcal{H}} = \Big\| \sqrt{\|E(k)\|} \Big\|_{L_2} = \left(\int \|E(k)\| \frac{dk}{(2\pi)^{d+1}} \right)^{1/2}.$$

The corollary now follows from Proposition 1.34 and Problem 1.26. □

CHAPTER 2

Fermionic Expansions

This chapter concerns an expansion that can be used to exhibit analytic control over the right-hand side of (1.7) in fermionic quantum field theory models, when the covariance S is "really nice." It is also used as one ingredient in a renormalization group procedure that controls the right-hand side of (1.8) when the covariance S is not so nice.

2.1. Notation and Definitions

Here are the main notations that we shall use throughout this chapter. Let

- $\mathbf{A}$ be the Grassmann algebra generated by $\{a_1, \ldots, a_D\}$. Think of $\{a_1, \ldots, a_D\}$ as some finite approximation to the set $\{\psi_{x,\sigma}, \bar{\psi}_{x,\sigma} \mid x \in \mathbb{R}^{d+1}, \sigma \in \{\uparrow, \downarrow\}\}$ of fields integrated out in a renormalization group transformation like
$$W(\psi, \bar{\psi}) \to \widetilde{W}(\Psi, \bar{\Psi}) = \log \frac{1}{Z} \int e^{W(\psi+\Psi, \bar{\psi}+\bar{\Psi})}\, d\mu_S(\psi, \bar{\psi}).$$
- $\mathbf{C}$ be the Grassmann algebra generated by $\{c_1, \ldots, c_D\}$. Think of $\{c_1, \ldots, c_D\}$ as some finite approximation to the set $\{\Psi_{x,\sigma}, \bar{\Psi}_{x,\sigma} \mid x \in \mathbb{R}^{d+1}, \sigma \in \{\uparrow, \downarrow\}\}$ of fields that are arguments of the output of the renormalization group transformation.
- $\mathbf{AC}$ be the Grassmann algebra generated by $\{a_1, \ldots, a_D, c_1, \ldots, c_D\}$.
- $S = (S_{ij})$ be a skew symmetric matrix of order D. Think of S as the "single scale" covariance $S^{(j)}$ of the Gaussian measure that is integrated out in the renormalization group step. See (1.8).
- $\int \cdot\, d\mu_S(a)$ be the Grassmann, Gaussian integral with covariance S. It is the unique linear map from $\mathbf{AC}$ to $\mathbf{C}$ satisfying
$$\int e^{\sum c_i a_i}\, d\mu_S(a) = e^{-(\sum c_i S_{ij} c_j)/2}.$$
In particular
$$\int a_i a_j\, d\mu_S(a) = S_{i,j}.$$
- $\mathcal{M}_r = \{(i_1, \ldots, i_r) \mid 1 \le i_1, \ldots, i_r \le D\}$ be the set of all multi indices of degree $r \ge 0$. For each $I \in \mathcal{M}_r$ set $a_I = a_{i_1} \cdots a_{i_r}$. By convention, $a_\emptyset = 1$.
- the space $(\mathbf{AC})^0$ of "interactions" is the linear subspace of $\mathbf{AC}$ of even Grassmann polynomials with no constant term. That is, polynomials of the form
$$W(c, a) = \sum_{\substack{l,r \in \mathbb{N} \\ 1 \le l+r \in 2\mathbb{Z}}} \sum_{\substack{L \in \mathcal{M}_l \\ J \in \mathcal{M}_r}} w_{l,r}(\mathrm{L}, \mathrm{J}) c_{\mathrm{L}} a_{\mathrm{J}}.$$

Usually, in the renormalization group map, the interaction is of the form $W(c+a)$. (See Definition 1.22). We do not require this.

Here are the main objects that shall concern us in this chapter.

DEFINITION 2.1. (a) The *renormalization group map* $\Omega\colon (\mathbf{AC})^0 \to \mathbf{C}^0$ is

$$\Omega(W)(c) = \log \frac{1}{Z_{W,S}} \int e^{W(c,a)}\, d\mu_S(a), \quad \text{where } Z_{W,S} = \int e^{W(0,a)}\, d\mu_S(a).$$

It is defined for all W's obeying $\int e^{W(0,a)}\, d\mu_S(a) \neq 0$. The factor $1/Z_{W,S}$ ensures that $\Omega(W)(0) = 0$, i.e., that $\Omega(W)(c)$ contains no constant term. Since $\Omega(W) = 0$ for $W = 0$

$$\begin{aligned}
\Omega(W)(c) &= \int_0^1 \frac{d}{d\epsilon}\Omega(\epsilon W)(c)\, d\epsilon \\
&= \int_0^1 \frac{\int W(c,a) e^{\epsilon W(c,a)}\, d\mu_S(a)}{\int e^{\epsilon W(c,a)}\, d\mu_S(a)}\, d\epsilon \\
&\qquad - \int_0^1 \frac{\int W(0,a) e^{\epsilon W(0,a)}\, d\mu_S(a)}{\int e^{\epsilon W(0,a)}\, d\mu_S(a)}\, d\epsilon.
\end{aligned} \tag{2.1}$$

Thus to get bounds on the renormalization group map, it suffices to get bounds on

(b) the *Schwinger functional* $\mathcal{S}\colon \mathbf{AC} \to \mathbf{C}$, defined by

$$S(f) = \frac{1}{\mathcal{Z}(c)} \int f(c,a) e^{W(c,a)}\, d\mu_S(a),$$

where $\mathcal{Z}(c) = \int e^{W(c,a)}\, d\mu_S(a)$. Despite our notation, $\mathcal{S}(f)$ is a function of W and S as well as f.

(c) Define the linear map $\mathrm{R}\colon \mathbf{AC} \to \mathbf{AC}$ by

$$\mathrm{R}(f)(c,a) = \int {:}e^{W(c,a+b)-W(c,a)} - 1{:}_b\, f(c,b)\, d\mu_S(b)$$

where $:\cdot:_b$ denotes Wick ordering of the b-field (see Section 1.6) and is determined by

$$:e^{\sum A_i a_i + \sum B_i b_i + \sum C_i c_i}:_b = e^{(\sum B_i S_{ij} B_j)/2} e^{\sum A_i a_i + \sum B_i b_i + \sum C_i c_i}$$

where $\{a_i\}$, $\{A_i\}$, $\{b_i\}$, $\{B_i\}$, $\{c_i\}$, $\{C_i\}$ are bases of six isomorphic vector spaces.

If you don't know what a Feynman diagram is, skip this paragraph. Diagrammatically, $\left(\mathcal{Z}(c)\right)^{-1} \int e^{W(c,a)} f(c,a)\, d\mu_S(a)$ is the sum of all connected (f is viewed as single, connected, vertex) Feynman diagrams with one f-vertex and arbitrary numbers of W-vertices and S-lines. The operation $\mathrm{R}(f)$ builds parts of those diagrams. It introduces those W-vertices that are connected directly to the f-vertex (i.e., that share a common S-line with f) and it introduces those lines that connect f either to itself or to a W-vertex.

PROBLEM 2.1. Let

$$F(a) = \sum_{j_i, j_2=1}^{D} f(j_1, j_2) a_{j_1} a_{j_2}$$

$$W(a) = \sum_{j_1, j_2, j_3, j_4=1}^{D} w(j_1, j_2, j_3, j_4) a_{j_1} a_{j_2} a_{j_3} a_{j_4}$$

with $f(j_1, j_2)$ and $w(j_1, j_2, j_3, j_4)$ antisymmetric under permutation of their arguments.

(a) Set

$$\mathcal{S}(\lambda) = \frac{1}{\mathcal{Z}_\lambda} \int F(a) e^{\lambda W(a)}\, d\mu_S(a), \quad \text{where } \mathcal{Z}_\lambda = \int e^{\lambda W(a)}\, d\mu_S(a).$$

Compute $d^\ell \mathcal{S}(\lambda)/d\lambda^\ell|_{\lambda=0}$ for $\ell = 0, 1, 2$.

(b) Set

$$\mathrm{R}(\lambda) = \int :e^{\lambda W(a+b) - \lambda W(a)} - 1:_b F(b)\, d\mu_S(b).$$

Compute $d^\ell\, \mathrm{R}(\lambda)/d\lambda^\ell|_{\lambda=0}$ for all $\ell \in \mathbb{N}$.

2.2. The Expansion—Algebra

To obtain the expansion that will be discussed in this chapter, expand the $(\mathbb{1} - \mathrm{R})^{-1}$ of the following Theorem, which shall be proven shortly, in a power series in R.

THEOREM 2.2. *Suppose that $W \in \mathbf{AC}^0$ is such that the kernel of $\mathbb{1} - \mathrm{R}$ is trivial. Then, for all f in* **AC**,

$$\mathcal{S}(f) = \int (\mathbb{1} - \mathrm{R})^{-1}(f)\, d\mu_S(a).$$

PROPOSITION 2.3. *For all f in* **AC** *and $W \in \mathbf{AC}^0$,*

$$\int f(c,a) e^{W(c,a)}\, d\mu_S(a) = \int f(c,b)\, d\mu_S(b) \int e^{W(c,a)}\, d\mu_S(a) + \int \mathrm{R}(f)(c,a) e^{W(c,a)}\, d\mu_S(a).$$

PROOF. Subbing in the definition of $\mathrm{R}(f)$,

$$\begin{aligned}\int f(c,b)\, d\mu_S(b) \int e^{W(c,a)}\, d\mu_S(a) &+ \int \mathrm{R}(f)(c,a) e^{W(c,a)}\, d\mu_S(a) \\ &= \int \left[\int :e^{W(c,a+b) - W(c,a)}:_b f(c,b)\, d\mu_S(b) \right] e^{W(c,a)}\, d\mu_S(a) \\ &= \int \int :e^{W(c,a+b)}:_b\, f(c,b)\, d\mu_S(b)\, d\mu_S(a)\end{aligned}$$

since $:e^{W(c,a+b) - W(c,a)}:_b = :e^{W(c,a+b)}:_b\, e^{-W(c,a)}$. Continuing,

$$\begin{aligned}\int f(c,b)\, d\mu_S(b) \int e^{W(c,a)}\, d\mu_S(a) &+ \int \mathrm{R}(f)(c,a) e^{W(c,a)}\, d\mu_S(a) \\ &= \iint :e^{W(c,a+b)}:_a\, f(c,b)\, d\mu_S(b)\, d\mu_S(a) \quad \text{by Problem 1.20} \\ &= \int f(c,b) e^{W(c,b)}\, d\mu_S(b) \quad \text{by Problems 1.11, 1.22} \\ &= \int f(c,a) e^{W(c,a)}\, d\mu_S(a).\end{aligned}$$

□

PROOF OF THEOREM 2.2. For all $g(c,a) \in \mathbf{AC}$

$$\int (\mathbb{1} - \mathrm{R})(g) e^{W(c,a)}\, d\mu_S(a) = \mathcal{Z}(c) \int g(c,a)\, d\mu_S(a)$$

by Proposition 2.3. If the kernel of $\mathbb{1} - \mathrm{R}$ is trivial, then we may choose $g = (\mathbb{1} - \mathrm{R})^{-1}(f)$. So

$$\int f(c,a)e^{W(c,a)}\,d\mu_S(a) = \mathcal{Z}(c)\int (\mathbb{1} - \mathrm{R})^{-1}(f)(c,a)\,d\mu_S(a).$$

The left-hand side does not vanish for all $f \in \mathbf{AC}$ (for example, for $f = e^{-W}$) so $\mathcal{Z}(c)$ is nonzero and

$$\frac{1}{\mathcal{Z}(c)}\int f(c,a)e^{W(c,a)}\,d\mu_S(a) = \int (\mathbb{1} - R)^{-1}(f)(c,a)\,d\mu_S(a). \qquad \square$$

2.3. The Expansion—Bounds

DEFINITION 2.4 (Norms). For any function $f\colon \mathcal{M}_r \to \mathbb{C}$, define

$$\|f\| = \max_{1\le i\le r}\max_{1\le k\le D}\sum_{\substack{j\in\mathcal{M}_r\\ j_i=k}} |f(\mathrm{J})|$$

$$|||f||| = \sum_{J\in\mathcal{M}_r} |f(\mathrm{J})|.$$

The norm $\|\cdot\|$, which is an "L^1 norm with one argument held fixed," is appropriate for kernels, like those appearing in interactions, that become translation invariant when the cutoffs are removed. Any $f(c,a) \in \mathbf{AC}$ has a unique representation

$$f(c,a) = \sum_{l,r\ge 0}\sum_{\substack{k_1,\dots,k_l\\ j_1,\dots,j_r}} f_{l,r}(k_1,\dots,k_l,\dots,j_r)c_{k_1}\cdots c_{k_l}a_{j_1}\cdots a_{j_r}$$

with each kernel $f_{l,r}(k_1,\dots,k_l,j_1,\dots,j_r)$ antisymmetric underseparate permutation of its k arguments and its j arguments. Define

$$\|f(c,a)\|_\alpha = \sum_{l,r\ge 0}\alpha^{l+r}\|f_{l,r}\|$$

$$|||f(c,a)|||_\alpha = \sum_{l,r\ge 0}\alpha^{l+r}|||f_{l,r}|||.$$

In the norms $\|f_{l,r}\|$ and $|||f_{l,r}|||$ on the right-hand side, $f_{l,r}$ is viewed as a function from $\mathcal{M}_{l+r}$ to $\mathbb{C}$.

PROBLEM 2.2. Define, for all $f\colon \mathcal{M}_r \to \mathbb{C}$ and $g\colon \mathcal{M}_s \to \mathbb{C}$ with $r,\ s \ge 1$ and $r + s > 2$, $f * g\colon \mathcal{M}_{r+s-2} \to \mathbb{C}$ by

$$f * g(j_1,\dots,j_{r+s-2}) = \sum_{k=1}^{D} f(j_1,\dots,j_{r-1},k)g(k,j_r,\dots,j_{r+s-2}).$$

Prove that

$$\|f * g\| \le \|f\|\,\|g\| \quad\text{and}\quad |||f * g||| \le \min\{|||f|||\,\|g\|, \|f\|\,|||g|||\}.$$

DEFINITION 2.5 (Hypotheses). We denote by

(HG) $$\left|\int b_{\mathrm{H}}{:}b_{\mathrm{J}}{:}\,d\mu_S(b)\right| \le \mathrm{F}^{|\mathrm{H}|+|\mathrm{J}|} \quad\text{for all } \mathrm{H},\mathrm{J} \in \bigcup_{r\le 0}\mathcal{M}_r$$

(HS) $$\|S\| \le \mathrm{F}^2\mathrm{D}$$

the main hypotheses on S that we shall use. So F is a measure of the "typical size of fields b in the support of the measure $d\mu_S(b)$" and D is a measure of the decay rate of S. Hypothesis (HG) is typically verified using Proposition 1.34 or Corollary 1.35. Hypotheses (HS) is typically verified using the techniques of Appendix C.

PROBLEM 2.3. Let $:\cdot:_S$ denote Wick ordering with respect to the covariance S.

(a) Prove that if

$$\left|\int b_{\rm H}{:}b_{\rm J}{:}_S\, d\mu_S(b)\right| \le {\rm F}^{|{\rm H}|+|{\rm J}|} \quad \text{for all } {\rm H},{\rm J} \in \bigcup_{r\ge 0} \mathcal{M}_r$$

then

$$\left|\int b_{\rm H}{:}b_{\rm J}{:}_{zS}\, d\mu_{zS}(b)\right| \le \left(\sqrt{|z|}\,{\rm F}\right)^{|{\rm H}|+|{\rm J}|} \quad \text{for all } {\rm H},{\rm J} \in \bigcup_{r\ge 0} \mathcal{M}_r.$$

Hint: first prove that $\int b_{\rm H}{:}b_{\rm J}{:}_{zS}\, d\mu_{zS}(b) = z^{(|{\rm H}|+|{\rm J}|)/2}\int b_{\rm H}{:}b_{\rm J}{:}_S\, d\mu_S(b)$.

(b) Prove that if

$$\left|\int b_{\rm H}{:}b_{\rm J}{:}_S\, d\mu_S(b)\right| \le {\rm F}^{|{\rm H}|+|{\rm J}|} \quad \text{and} \quad \left|\int b_{\rm H}{:}b_{\rm J}{:}_T\, d\mu_T(b)\right| \le {\rm G}^{|{\rm H}|+|{\rm J}|}$$

for all $\rm H, J \in \bigcup_{r\ge 0}\mathcal{M}_r$, then

$$\left|\int b_{\rm H}{:}b_{\rm J}{:}_{S+T}\, d\mu_{S+T}(b)\right| \le ({\rm F}+{\rm G})^{|{\rm H}|+|{\rm J}|} \quad \text{for all } {\rm H},{\rm J} \in \bigcup_{r\ge 0} \mathcal{M}_r.$$

Hint: Proposition 1.21 and Problem 1.21.

THEOREM 2.6. *Assume Hypotheses* (HG) *and* (HS). *Let* $\alpha \ge 2$ *and* $W \in \mathbf{AC}^0$ *obey* $D\|W\|_{(\alpha+1)\rm F} \le \frac{1}{3}$. *Then, for all* $f \in \mathbf{AC}$,

$$\|\,{\rm R}(f)\|_{\alpha{\rm F}} \le \frac{3}{\alpha} D\|W\|_{(\alpha+1)F}\|f\|_{\alpha{\rm F}}$$
$$|||{\rm R}(f)|||_{\alpha{\rm F}} \le \frac{3}{\alpha} D\|W\|_{(\alpha+1){\rm F}}|||f|||_{\alpha{\rm F}}.$$

The proof of this theorem follows that of Lemma 2.12.

COROLLARY 2.7. *Assume Hypotheses* (HG) *and* (HS). *Let* $\alpha \ge 2$ *and* $W \in \mathbf{AC}^0$ *obey* $D\|W\|_{(\alpha+1)F} \le \frac{1}{3}$. *Then for all* $f \in \mathbf{AC}$,

$$\|\mathcal{S}(f)(c) - \mathcal{S}(f)(0)\|_{\alpha{\rm F}} \le \frac{\alpha}{\alpha-1}\|f\|_{\alpha{\rm F}}$$
$$|||\mathcal{S}(f)|||_{\alpha{\rm F}} \le \frac{\alpha}{\alpha-1}|||f|||_{\alpha{\rm F}}$$
$$\|\Omega(W)\|_{\alpha{\rm F}} \le \frac{\alpha}{\alpha-1}\|W\|_{\alpha{\rm F}}.$$

The proof of this corollary follows that of Lemma 2.12.

Let

$$W(c,a) = \sum_{l,r\in\mathbb{N}} \sum_{\substack{{\rm L}\in\mathcal{M}_l \\ {\rm J}\in\mathcal{M}_r}} w_{l,r}({\rm L},{\rm J}) C_{\rm L} a_{\rm J}$$

where $w_{l,r}({\rm L},{\rm J})$ is a function which is separately antisymmetric under permutations of its L and J arguments and that vanishes identically when $l+r$ is zero or odd.

With this antisymmetry

$$W(c,a+b) - W(c,a) = \sum_{\substack{l,r\ge 0\\ s\ge 1}} \sum_{\substack{\mathrm{L}\in\mathcal{M}_l\\ \mathrm{J}\in\mathcal{M}_r\\ \mathrm{K}\in\mathcal{M}_s}} \binom{r+s}{s} w_{l,r+s}(\mathrm{L},\mathrm{J.K}) c_{\mathrm{L}} a_{\mathrm{J}} b_{\mathrm{K}}$$

where $\mathrm{J.K} = (j_1,\ldots,j_r,k_1,\ldots,k_s)$ when $\mathrm{J} = (j_1,\ldots,j_r)$ and $\mathrm{K} = (k_1,\ldots,k_s)$. That is, J.K is the concatenation of J and K. So

$$:e^{W(c,a+b)-W(c,a)} - 1:_b =$$

$$\sum_{\ell>0} \frac{l}{\ell!} \sum_{\substack{l_i,r_i\ge 0\\ s_i\ge 1}} \sum_{\substack{\mathrm{L}_i\in\mathcal{M}_{l_i}\\ \mathrm{J}_i\in\mathcal{M}_{r_i}\\ \mathrm{K}_i\in\mathcal{M}_{s_i}}} :\prod_{i=1}^{l} \binom{r_i+s_i}{s_i} w_{l_i,r_i+s_i}(\mathrm{L}_i,\mathrm{J}_i.\mathrm{K}_i) c_{\mathrm{L}_i} a_{\mathrm{J}_i} b_{\mathrm{K}_i}:_b$$

with the index i in the second and third sums running from 1 to ℓ. Hence

$$\mathrm{R}(f) = \sum_{\ell>0} \sum_{\substack{\mathbf{r},\mathbf{s},\mathbf{l}\in\mathbb{N}^\ell\\ s_i\ge 1}} \frac{1}{\ell!} \prod_{i=1}^{\ell} \binom{r_i+s_i}{s_i} \mathrm{R}^{\mathbf{s}}_{\mathbf{l},\mathbf{r}}(f),$$

where

$$\mathrm{R}^{\mathbf{s}}_{\mathbf{l},\mathbf{r}}(f) = \sum_{\substack{\mathrm{L}_i\in\mathcal{M}_{l_i}\\ \mathrm{J}_i\in\mathcal{M}_{r_i}\\ \mathrm{K}_i\in\mathcal{M}_{s_i}}} \int :\prod_{i=1}^{\ell} w_{l_i r_i+s_i}(\mathrm{L}_i,\mathrm{J}_i.\mathrm{K}_i) c_{\mathrm{L}_i} a_{\mathrm{J}_i} b_{\mathrm{K}_i}:_b f(c,b)\, d\mu_S(b).$$

PROBLEM 2.4. Let $W(a) = \sum_{i,j=1}^{D} w(i,j) a_i a_j$ with $w(i,j) = -w(j,i)$. Verify that

$$W(a+b) - W(a) = \sum_{i,j=1}^{D} w(i,j) b_i b_j + 2\sum_{i,j=1}^{D} w(i,j) a_i b_j.$$

PROBLEM 2.5. Assume Hypothesis (HG). Let $s, s', m \ge 1$ and

$$f(b) = \sum_{\mathrm{H}\in\mathcal{M}_m} f_m(\mathrm{H}) b_{\mathrm{H}}, \qquad W(b) = \sum_{\mathrm{K}\in\mathcal{M}_s} w_s(\mathrm{K}) b_{\mathrm{K}}, \qquad W'(b) = \sum_{\mathrm{K}'\in\mathcal{M}_{s'}} w'_{s'}(\mathrm{K}') b_{\mathrm{K}'}.$$

(a) Prove that

$$\left| \int :W(b):_b f(b)\, d\mu_S(b) \right| \le m \mathrm{F}^{m+s-2} |||f_m|||\, \|S\|\, \|w_s\|.$$

(b) Prove that $\int :W(b)W'(b):_b f(b)\, d\mu_S(b) = 0$ if $m = 1$ and

$$\left| \int :W(b)W'(b):_b f(b)\, d\mu_S(b) \right| \le m(m-1) \mathrm{F}^{m+s+s'-4} |||f_m|||\, \|S\|^2 \|w_s\|\, \|w'_s\|$$

if $m \ge 2$.

PROPOSITION 2.8. *Assume Hypothesis* (HG). *Let*

$$f^{(p,m)}(c,a) = \sum_{\substack{\mathrm{H}\in\mathcal{M}_m\\ \mathrm{I}\in\mathcal{M}_p}} f_{p,m}(\mathrm{I},\mathrm{H}) c_{\mathrm{I}} a_{\mathrm{H}}.$$

Let $\mathbf{r}, \mathbf{s}, \mathbf{l} \in \mathbb{N}^\ell$ *with each* $s_i \geq 1$. *If* $m < \ell$, $R^{\mathbf{s}}_{\mathbf{l},\mathbf{r}}(f^{(p,m)})$ *vanishes. If* $m \geq \ell$

$$\| \mathrm{R}^{\mathbf{s}}_{\mathbf{l},\mathbf{r}}(f^{(p,m)})\|_1 \leq \ell! \binom{m}{\ell} \mathrm{F}^m \|f_{p,m}\| \prod_{i=1}^{\ell} (\|S\|\mathrm{F}^{s_i-2}\|w_{l,r_i+s_i}\|)$$

$$|||\mathrm{R}^{\mathbf{s}}_{\mathbf{l},\mathbf{r}}(f^{(p,m)})|||_1 \leq \ell! \binom{m}{\ell} \mathrm{F}^m |||f_{p,m}||| \prod_{i=1}^{\ell} (\|S\|\mathrm{F}^{s_i-2}\|w_{l_i,r_i+s_i}\|).$$

PROOF. We have

$$\mathrm{R}^{\mathbf{s}}_{\mathbf{l},\mathbf{r}}\big(f^{(p,m)}\big) = \pm \sum_{\mathrm{I}\in\mathcal{M}_p} \sum_{\substack{1\leq i\leq \ell \\ \mathrm{J}_i\in\mathcal{M}_{r_i} \\ \mathrm{L}_i\in\mathcal{M}_{l_i}}} f_{\mathbf{l},\mathbf{r},\mathbf{s}}(\mathrm{L}_1,\ldots,\mathrm{L}_\ell,\mathrm{I},\mathrm{J}_1,\ldots,\mathrm{J}_\ell) c_{\mathrm{L}_1}\cdots c_{\mathrm{L}_\ell} c_{\mathrm{I}} a_{\mathrm{J}_1}\cdots a_{\mathrm{J}_\ell}$$

with

$$f_{\mathbf{l},\mathbf{r},\mathbf{s}}(\mathrm{L}_1,\ldots,\mathrm{L}_\ell,\mathrm{I},\mathrm{J}_1,\ldots,\mathrm{J}_\ell) = \sum_{\substack{\mathrm{H}\in\mathcal{M}_m \\ \mathrm{K}_i\in\mathcal{M}_{s_i}}} \int {:}\prod_{i=1}^{\ell} w_{l_i,r_i+s_i}(\mathrm{L}_i,\mathrm{J}_i,\mathrm{K}_i) b_{\mathrm{K}_i}{:}_b f_{p,m}(\mathrm{I},\mathrm{H}) b_{\mathrm{H}}\, d\mu_S(b).$$

The integral over b is bounded in Lemma 2.9 below. It shows that

$$f_{\mathbf{l},\mathbf{r},\mathbf{s}}(L_1,\ldots,L_\ell,\mathrm{I},\mathrm{J}_1,\ldots,\mathrm{J}_\ell)$$

vanishes if $m < \ell$ and is, for $m \geq \ell$, bounded by

$$|f_{\mathbf{l},\mathbf{r},\mathbf{s}}(\mathrm{L}_1,\ldots,\mathrm{L}_\ell,\mathrm{I},\mathrm{J}_1,\ldots,\mathrm{J}_\ell| \leq \ell! \binom{m}{\ell} T(\mathrm{L}_1,\ldots,\mathrm{L}_\ell,\mathrm{I},\mathrm{J}_1,\ldots,\mathrm{J}_\ell)\mathrm{F}^{m+\sum s_i - 2\ell}$$

where

$$T(\mathrm{L}_1,\ldots,\mathrm{L}_\ell,\mathrm{I},J_1,\ldots,\mathrm{J}_\ell) = \sum_{\mathrm{H}\in\mathcal{M}_m} \prod_{i=1}^{\ell} \Big(\sum_{k_i=1}^{D} |u_i(\mathrm{L}_i,\mathrm{J}_i,k_i)|\,|S_{k_i,h_i}| \Big) |f_{p,m}(\mathrm{I},\mathrm{H})|$$

and, for each $i = 1,\ldots,\ell$

$$u_i(\mathrm{L}_i,\mathrm{J}_i,k_i) = \sum_{\widetilde{\mathrm{K}}_i\in\mathcal{M}_{s_i-1}} \big|w_{l_i,r_i+s_i}\big(\mathrm{L}_i,\mathrm{J}_i,(k_i)\cdot\widetilde{\mathrm{K}}_i\big)\big|.$$

Recall that $(k_1).\widetilde{\mathrm{K}} = (k_1,\tilde{k}_2,\ldots,\tilde{k}_s)$ when $\widetilde{\mathrm{K}} = (\tilde{k}_2,\ldots,\tilde{k}_s)$. By construction $\|u_i\| = \|w_{l_i,r_i+s_i}\|$. By Lemma 2.10, below

$$\|T\| \leq \|f_{p,m}\|\,\|S\|^\ell \prod_{i=1}^{\ell} \|u_i\| \leq \|f_{p,m}\|\,\|S\|^\ell \prod_{i=1}^{\ell} \|w_{l_i,r_i+s_i}\|$$

and hence

$$\|f_{\mathbf{l},\mathbf{r},\mathbf{s}}\| \leq \ell! \binom{m}{\ell} \mathrm{F}^{m+\sum s_i-2\ell} \|f_{p,m}\|\|S\|^\ell \prod_{i=1}^{\ell} \|w_{l_i,r_i+s_i}\|.$$

Similarly, the second bound follows from $|||T||| \leq |||f_{p,m}|||\,\|S\|^\ell \prod_{i=1}^{\ell} \|u_i\|$. □

LEMMA 2.9. *Assume Hypothesis* (HG). *Then*

$$\Big|\int {:}\prod_{i=1}^{\ell} b_{\mathrm{K}_i}{:} b_{\mathrm{H}}\, d\mu_S(b)\Big| \leq \mathrm{F}^{|\mathrm{H}|+\sum|\mathrm{K}_i|-2\ell} \sum_{\substack{\mu_1,\ldots,\mu_\ell=1 \\ \text{all different}}}^{|\mathrm{H}|} \prod_{i=1}^{\ell} |S_{k_{i1},h_{\mu_i}}|.$$

PROOF. For convenience, set $j_i = k_{i1}$ and $\widetilde{K}_i = K_i \setminus \{k_{i1}\}$ for each $i = 1, \ldots, \ell$. By antisymmetry,

$$\int :\prod_{i=1}^{\ell} b_{K_i}: b_H \, d\mu_S(b) = \pm \int :\prod_{i=1}^{\ell} b_{\widetilde{K}_i} b_{j_1} \ldots b_{j_\ell}: b_H \, d\mu_S(b).$$

Recall the integration by parts formula (Problem 1.19)

$$\int :b_K b_j: f(b) \, d\mu_S(b) = \sum_{m=1}^{D} S_{j,m} \int :b_K: \frac{\partial}{\partial b_m} f(b) \, d\mu_S(b)$$

and the definition (Definition 1.11)

$$\frac{\partial}{\partial b_m} b_H = \begin{cases} 0, & \text{if } m \notin H; \\ (-1)^{|J|} b_J b_K, & \text{if } b_H = b_J b_m b_K \end{cases}.$$

of left partial derivative. Integrate by parts successively with respect to $b_{j_\ell} \cdots b_{j_1}$

$$\int :\prod_{i=1}^{\ell} b_{K_i}: b_H \, d\mu_S(b) = \pm \int :\prod_{i=1}^{\ell} b_{\widetilde{K}_i}: \left[\prod_{i=1}^{\ell} \Big(\sum_{m=1}^{D} S_{j_i,m} \frac{\partial}{\partial b_m} \Big) b_H \right] d\mu_S(b)$$

and then apply Leibniz's rule

$$\prod_{i=1}^{\ell} \Big(\sum_{m=1}^{D} S_{j_i,m} \frac{\partial}{\partial b_m} \Big) b_H = \sum_{\substack{\mu_1,\ldots,\mu_\ell=1 \\ \text{all different}}}^{|H|} \pm \Big(\prod_{i=1}^{\ell} S_{j_i,h_{\mu_i}} \Big) b_{H \setminus \{h_{\mu_1},\ldots,h_{\mu_\ell}\}}$$

and Hypothesis (HG). □

LEMMA 2.10. *Let*

$$T(J_1, \ldots, J_\ell, I) = \sum_{H \in \mathcal{M}_m} \prod_{i=1}^{\ell} \Big(\sum_{k_i=1}^{D} |u_i(J_i, k_i)| |S_{k_i,h_i}| \Big) |f_{p,m}(I, H)|$$

with $\ell \le m$. Then

$$\|T\| \le \|f_{p,m}\| \, \|S\|^\ell \prod_{i=1}^{\ell} \|u_i\|$$

$$|||T||| \le |||f_{p,m}||| \, \|S\|^\ell \prod_{i=1}^{\ell} \|u_i\|.$$

PROOF. For the triple norm,

$$\begin{aligned} |||T||| &= \sum_{\substack{J_i \in \mathcal{M}_{r_i}, 1 \le i \le \ell \\ I \in \mathcal{M}_p}} T(J_1, \ldots, J_\ell, I) \\ &= \sum_{\substack{J_i \in \mathcal{M}_{r_i}, 1 \le i \le \ell \\ I \in \mathcal{M}_p, H \in \mathcal{M}_m}} \prod_{i=1}^{\ell} \Big(\sum_{k_i=1}^{D} |u_i(J_i, k_i)| |S_{k_i,h_i}| \Big) |f_{p,m}(I, H)| \\ &= \sum_{\substack{I \in \mathcal{M}_p \\ H \in \mathcal{M}_m}} \prod_{i=1}^{\ell} \Big(\sum_{k_i} \sum_{J_i \in \mathcal{M}_{r_i}} |u_i(J_i, k_i)| |S_{k_i,h_i}| \Big) |f_{p,m}(I, H)| \end{aligned}$$

$$= \sum_{\substack{\mathrm{I}\in\mathcal{M}_p\\ \mathrm{H}\in\mathcal{M}_m}} \Big(\prod_{i=1}^{\ell} v_i(h_i)\Big)|f_{p,m}(\mathrm{I},\mathrm{H})|$$

where

$$v_i(h_i) = \sum_{k_i}\sum_{\mathrm{J}_i\in\mathcal{M}_{r_i}} |u_i(\mathrm{J}_i,k_i)||S_{k_i,h_i}|.$$

Since

$$\begin{aligned}\sup_{h_i} v_i(h_i) &= \sup_{h_i}\sum_{k_i}\Big(\sum_{\mathrm{J}_i\in\mathcal{M}_{r_i}} |u_i(\mathrm{J}_i,k_i)|\Big)|S_{k_i,h_i}|\\ &\le \sup_{k_i}\Big(\sum_{J_i\in\mathcal{M}_{r_i}} |u_i(\mathrm{J}_i,k_i)|\Big)\sup_{h_i}\sum_{k_i}|S_{k_i,h_i}|\\ &\le \|u_i\|\|S\|\end{aligned}$$

we have

$$|||T||| \le \sum_{\substack{\mathrm{I}\in\mathcal{M}_p\\ \mathrm{H}\in\mathcal{M}_m}} \Big(\prod_{i=1}^{\ell}\|u_i\|\,\|S\|\Big)|f_{p,m}(\mathrm{I},\mathrm{H})| = |||f_{p,m}|||\prod_{i=1}^{\ell}(\|u_i\|\,\|S\|).$$

The proof for the double norm, i.e., the norm with one external argument sup'd over rather than summed over, is similar. There are two cases. Either the external argument that is being sup'd over is a component of I or it is a component of one of the J_j's. In the former case,

$$\begin{aligned}&\sup_{1\le i_1\le D}\sum_{\substack{\mathrm{J}_i\in\mathcal{M}_{r_i},1\le i\le\ell\\ \tilde{\mathrm{I}}\in\mathcal{M}_{p-1},\mathrm{H}\in\mathcal{M}_m}}\prod_{i=1}^{\ell}\Big(\sum_{k_i=1}^{D}|u_i(\mathrm{J}_i,k_i)|\,|S_{k_i,h_i}|\Big)\big|f_{p,m}\big((i_1)\tilde{\mathrm{I}},\mathrm{H}\big)\big|\\ &= \sup_{1\le i_1\le D}\sum_{\substack{\tilde{\mathrm{I}}\in\mathcal{M}_{p-1}\\ \mathrm{H}\in\mathcal{M}_m}}\Big(\prod_{i=1}^{\ell}v_i(h_i)\Big)\big|f_{p,m}\big((i_1)\tilde{\mathrm{I}},\mathrm{H}\big)\big|\\ &\le \sup_{1\le i_1\le D}\sum_{\substack{\tilde{\mathrm{I}}\in\mathcal{M}_{p-1}\\ \mathrm{H}\in\mathcal{M}_m}}\Big(\prod_{i=1}^{\ell}\|u_i\|\,\|S\|\Big)\big|f_{p,m}\big((i_1)\tilde{\mathrm{I}},\mathrm{H}\big)\big| \le \|f_{p,m}\|\prod_{i=1}^{\ell}(\|u_i\|\,\|S\|).\end{aligned}$$

In the latter case, if a component of, for example, J_1, is to be sup'd over

$$\begin{aligned}&\sup_{1\le j_1\le D}\sum_{\substack{\tilde{\mathrm{J}}_1\in\mathcal{M}_{r_1-1}\\ \mathrm{J}_i\in\mathcal{M}_{r_i},2\le i\le\ell\\ \mathrm{I}\in\mathcal{M}_p,\mathrm{H}\in\mathcal{M}_m}}\Big(\sum_{k_1=1}^{D}\big|u_1\big((j_1).\tilde{\mathrm{J}}_1,k_1\big)\big|\,|S_{k_1,h_1}|\Big)\\ &\qquad\times\prod_{i=2}^{\ell}\Big(\sum_{k_i=1}^{D}|u_i(\mathrm{J}_i,k_i)||S_{k_i,h_i}|\Big)|f_{p,m}(\mathrm{I},\mathrm{H})|\\ &= \sup_{1\le j_1\le D}\sum_{\substack{\tilde{\mathrm{J}}_1\in\mathcal{M}_{r_1-1}\\ \mathrm{I}\in\mathcal{M}_p,\mathrm{H}\in\mathcal{M}_m}}\Big(\sum_{k_1=1}^{D}\big|u_1\big((j_1).\tilde{\mathrm{J}}_1,k_1\big)\big|\,|S_{k_1,h_1}|\Big)\Big(\prod_{i=2}^{\ell}v_i(h_i)\Big)|f_{p,m}(\mathrm{I},\mathrm{H})|\end{aligned}$$

$$\leq \left(\prod_{i=2}^{\ell} \|u_i\| \|S\|\right) \sup_{1\leq j_1\leq D} \sum_{\substack{\tilde{J}_1\in\mathcal{M}_{r_1-1}\\ I\in\mathcal{M}_p, H\in\mathcal{M}_m}} \sum_{k_1=1}^{D} \left|u_1\big((j_1).\tilde{J}_1, k_1\big)\right| |S_{k_1,h_1}| |f_{p,m}(I,H)|$$

$$\leq \|f_{p,m}\| \prod_{i=1}^{\ell} (\|u_i\| \, \|S\|)$$

by Problem 2.2, twice. □

LEMMA 2.11. *Assume Hypotheses* (HG) *and* (HS). *Let*

$$f^{(p,m)}(c,a) = \sum_{\substack{I\in\mathcal{M}_p\\ H\in\mathcal{M}_m}} f_{p,m}(I,H) c_I a_H.$$

For all $\alpha \geq 2$ *and* $\ell \geq 1$

$$\sum_{\substack{\mathbf{r},\mathbf{s},\mathbf{l}\in\mathbb{N}^\ell\\ s_i\geq 1}} \frac{1}{\ell!} \prod_{i=1}^{\ell} \binom{r_i+s_i}{s_i} \| \mathrm{R}^{\mathbf{s}}_{\mathbf{l},\mathbf{r}}(f^{(p,m)})\|_{\alpha \mathrm{F}} \leq \frac{2}{\alpha} \|f^{(p,m)}\|_{\alpha \mathrm{F}} [\mathrm{D}\|W\|_{(\alpha+1)\mathrm{F}}]^\ell$$

$$\sum_{\substack{\mathbf{r},\mathbf{s},\mathbf{l}\in\mathbb{N}^\ell\\ s_i\geq 1}} \frac{1}{\ell!} \prod_{i=1}^{\ell} \binom{r_i+s_i}{s_i} |||\mathrm{R}^{\mathbf{s}}_{\mathbf{l},\mathbf{r}}(f^{(p,m)})|||_{\alpha \mathrm{F}} \leq \frac{2}{\alpha} |||f^{(p,m)}|||_{\alpha \mathrm{F}} [\mathrm{D}\|W\|_{(\alpha+1)\mathrm{F}}]^\ell.$$

PROOF. We prove the bound with the $\|\cdot\|$ norm. The proof for $|||\cdot|||$ is identical. We may assume, without loss of generality, that $m \geq \ell$. By Proposition 2.8, since $\binom{m}{\ell} \leq 2^m$,

$$\begin{aligned}
\frac{1}{\ell!} \| \mathrm{R}^{\mathbf{s}}_{\mathbf{l},\mathbf{r}}(f^{(p,m)})\|_{\alpha \mathrm{F}} &\\
&= \frac{1}{\ell!} \alpha^{\sum l_i + \sum r_i + p} \mathrm{F}^{\sum l_i + \sum r_i + p} \|\mathbf{R}^{\mathbf{s}}_{\mathbf{l},\mathbf{r}}(f^{(p,m)})\|_1 \\
&\leq \alpha^{\sum l_i + \sum r_i + p} \mathrm{F}^{\sum l_i + \sum r_i + p} \binom{m}{\ell} \mathrm{F}^m \|f_{p,m}\| \prod_{i=1}^{\ell} (\|S\| \mathrm{F}^{s_i-2} \|w_{\ell_i, r_i+s_i}\|) \\
&\leq 2^m \alpha^p \mathrm{F}^{m+p} \|f_{p,m}\| \prod_{i=1}^{\ell} (\mathrm{D}\alpha^{l_i+r_i} \mathrm{F}^{l_i+r_i+s_i} \|w_{l_i,r_i+s_i}\|) \\
&= \left(\frac{2}{\alpha}\right)^m \|f_{p,m}\|_{\alpha \mathrm{F}} \prod_{i=1}^{\ell} (\mathrm{D}\alpha^{l_i+r_i} \mathrm{F}^{l_i+r_i+s_i} \|w_{l_i,r_i+s_i}\|).
\end{aligned}$$

As $\alpha \geq 2$ and $m \geq \ell \geq 1$,

$$\begin{aligned}
\sum_{\substack{\mathbf{r},\mathbf{s},\mathbf{l}\in\mathbb{N}^\ell\\ s_i\geq 1}} \frac{1}{\ell!} \prod_{i=1}^{\ell} \binom{r_i+s_i}{s_i} \| \mathrm{R}^{\mathbf{s}}_{\mathbf{l},\mathbf{r}}(f^{(p,m)})\|_{\alpha \mathrm{F}} &\\
&\leq \frac{2}{\alpha} \|f^{(p,m)}\|_{\alpha \mathrm{F}} \sum_{\substack{\mathbf{r},\mathbf{s},\mathbf{l}\in\mathbb{N}^\ell\\ s_i\geq 1}} \prod_{i=1}^{\ell} \left[\binom{r_i+s_i}{s_i} \mathrm{D}\alpha^{l_i+r_i} \mathrm{F}^{l_i+r_i+s_i} \|w_{l_i,r_i,+s_i}\|\right]
\end{aligned}$$

$$= \frac{2}{\alpha}\|f^{(p,m)}\|_{\alpha\mathrm{F}}\Big[\mathrm{D}\sum_{\substack{r,s,l\in\mathbb{N}\\ s\geq 1}}\binom{r+s}{s}\alpha^r\alpha^l\mathrm{F}^{l+r+s}\|w_{l,r+s}\|\Big]^\ell$$

$$= \frac{2}{\alpha}\|f^{(p,m)}\|_{\alpha\mathrm{F}}\Big[\mathrm{D}\sum_{l\in\mathbb{N}}\sum_{q\geq 1}\sum_{s=1}^{q}\binom{q}{s}\alpha^{q-s}\alpha^l\mathrm{F}^{l+q}\|w_{l,q}\|\Big]^\ell$$

$$\leq \frac{2}{\alpha}\|f^{(p,m)}\|_{\alpha\mathrm{F}}\Big[\mathrm{D}\sum_{q,l\in\mathbb{N}}(\alpha+1)^q\alpha^l\mathrm{F}^{l+q}\|w_{l,q}\|\Big]^\ell$$

$$\leq \frac{2}{\alpha}\|f^{(p,m)}\|_{\alpha\mathrm{F}}\big[\mathrm{D}\|W\|_{(\alpha+1)\mathrm{F}}\big]^\ell. \qquad \square$$

PROOF OF THEOREM 2.6.

$$\|\mathrm{R}(f)\|_{\alpha\mathrm{F}} \leq \sum_{\ell>0}\sum_{\substack{\mathbf{r},\mathbf{s},\mathbf{l}\in\mathbb{N}^\ell\\ s_i\geq 1}}\frac{1}{\ell!}\prod_{i=1}^{\ell}\binom{r_i+s_i}{s_i}\|\mathrm{R}^{\mathbf{s}}_{\mathbf{l},\mathbf{r}}(f)\|_{\alpha\mathrm{F}}$$

$$\leq \sum_{\ell>0}\sum_{m,p}\frac{2}{\alpha}\|f^{(p,m)}\|_{\alpha\mathrm{F}}\big[\mathrm{D}\|W\|_{(\alpha+1)\mathrm{F}}\big]^\ell$$

$$= \frac{2}{\alpha}\|f\|_{\alpha\mathrm{F}}\frac{\mathrm{D}\|W\|_{(\alpha+1)\mathrm{F}}}{1-\mathrm{D}\|W\|_{(\alpha+1)\mathrm{F}}}$$

$$\leq \frac{3}{\alpha}\|f\|_{\alpha\mathrm{F}}\mathrm{D}\|W\|_{(\alpha+1)\mathrm{F}}.$$

The proof for the other norm is similar. $\square$

LEMMA 2.12. *Assume Hypothesis* (HG). *If $\alpha\geq 1$ then, for all $g(a,c)\in\mathbf{AC}$*

$$\Big|\Big|\Big|\int g(a,c)\,d\mu_S(a)\Big|\Big|\Big|_{\alpha\mathrm{F}} \leq |||g(a,c)|||_{\alpha\mathrm{F}}$$

$$\Big\|\int[g(a,c)-g(a,0)]\,d\mu_S(a)\Big\|_{\alpha\mathrm{F}} \leq \|g(a,c)\|_{\alpha\mathrm{F}}.$$

PROOF. Let

$$g(a,c) = \sum_{l,r\geq 0}\sum_{\substack{\mathrm{L}\in\mathcal{M}_l\\ \mathrm{J}\in\mathcal{M}_r}} g_{l,r}(\mathrm{L},\mathrm{J})c_{\mathrm{L}}a_{\mathrm{J}}$$

with $g_{l,r}(\mathrm{L},\mathrm{J})$ antisymmetric under separate permutations of its L and J arguments. Then

$$\Big|\Big|\Big|\int g(a,c)\,d\mu_S(a)\Big|\Big|\Big|_{\alpha\mathrm{F}} = \Big|\Big|\Big|\sum_{l,r\geq 0}\sum_{\substack{\mathrm{L}\in\mathcal{M}_l\\ \mathrm{J}\in\mathcal{M}_r}} g_{l,r}(\mathrm{L},\mathrm{J})c_{\mathrm{L}}\int a_{\mathrm{J}}\,d\mu_S(a)\Big|\Big|\Big|_{\alpha\mathrm{F}}$$

$$= \sum_{l\geq 0}\alpha^l\mathrm{F}^l\sum_{\mathrm{L}\in\mathcal{M}_l}\Big|\sum_{r\geq 0}\sum_{\mathrm{J}\in\mathcal{M}_r} g_{l,r}(\mathrm{L},\mathrm{J})\int a_{\mathrm{J}}\,d\mu_S(a)\Big|$$

$$\leq \sum_{l,r\geq 0}\alpha^l\mathrm{F}^{l+r}\sum_{\substack{\mathrm{L}\in\mathcal{M}_l\\ \mathrm{J}\in\mathcal{M}_r}}|g_{l,r}(\mathrm{L},\mathrm{J})|$$

$$\leq |||g(a,c)|||_{\alpha\mathrm{F}}.$$

Similarly,

$$\begin{aligned}
&\left\| \int [g(a,c) - g(a,0)\, d\mu_S(a) \right\|_{\alpha\mathrm{F}} \\
&\qquad = \left\| \sum_{\substack{r\ge 0\\ l\ge 1}} \sum_{\substack{\mathrm{L}\in\mathcal{M}_l\\ \mathrm{J}\in\mathcal{M}_r}} g_{l,r}(\mathrm{L},\mathrm{J}) c_{\mathrm{L}} \int a_{\mathrm{J}}\, d\mu_S(a) \right\|_{\alpha\mathrm{F}} \\
&\qquad = \sum_{l\ge 1} \alpha^l \mathrm{F}^l \sup_{1\le k\le n} \sum_{\tilde{\mathrm{L}}\in\mathcal{M}_{l-1}} \left| \sum_{r\ge 0} \sum_{\mathrm{J}\in\mathcal{M}_r} g_{l,r}\big((k)\tilde{\mathrm{L}},\mathrm{J}\big) \int a_{\mathrm{J}}\, d\mu_S(a) \right| \\
&\qquad \le \sum_{\substack{l\ge 1\\ r\ge 0}} \alpha^l \mathrm{F}^{l+r} \sup_{1\le k\le n} \sum_{\substack{\mathrm{L}\in\mathcal{M}_{l-1}\\ \mathrm{J}\in\mathcal{M}_r}} \left| g_{l,r}\big((k)\tilde{\mathrm{L}},\mathrm{J}\big) \right| \\
&\qquad \le \|g(a,c)\|_{\alpha\mathrm{F}}. \qquad \square
\end{aligned}$$

PROOF OF COROLLARY 2.7. Set $g = (\mathbb{1} - \mathrm{R})^{-1} f$. Then

$$\begin{aligned}
&\|\mathcal{S}(f)(c) - \mathcal{S}(f)(0)\|_{\alpha\mathrm{F}} \\
&\qquad = \int [g(a,c) - g(a,0)]\, d\mu_S(a)\|_{\alpha\mathrm{F}} && \text{(Theorem 2.2)} \\
&\qquad \le \|(\mathbb{1} - \mathrm{R})^{-1}(f)\|_{\alpha\mathrm{F}} && \text{(Lemma 2.12)} \\
&\qquad \le \frac{1}{1 - 3\mathrm{D}\|W\|_{(\alpha+1)\mathrm{F}}/\alpha} \|f\|_{\alpha\mathrm{F}} \le \frac{1}{1-1/\alpha}\|f\|_{\alpha\mathrm{F}} && \text{(Theorem 2.6).}
\end{aligned}$$

The argument for $|||\mathcal{S}(f)|||_{\alpha\mathrm{F}}$ is identical.

With the more detailed notation

$$\mathcal{S}(f,W) = \frac{\int f(a,c) e^{W(a,c)}\, d\mu_S(a)}{\int e^{W(a,c)}\, d\mu_S(a)}$$

we have, by (2.1) followed by the first bound of this corollary, with the replacements $W \to \varepsilon W$ and $f \to W$,

$$\begin{aligned}
\|\Omega(w)\|_{\alpha\mathrm{F}} &= \left\| \int_0^1 [\mathcal{S}(W,\varepsilon W)(c) - \mathcal{S}(W,\varepsilon W)(0)]\, d\varepsilon \right\|_{\alpha\mathrm{F}} \\
&\le \int_0^1 \frac{\alpha}{\alpha-1} \|W\|_{\alpha\mathrm{F}}\, d\varepsilon = \frac{\alpha}{\alpha-1}\|W\|_{\alpha\mathrm{F}}. \qquad \square
\end{aligned}$$

We end this subsection with a proof of the continuity of the Schwinger functional

$$\mathcal{S}(f; W, S) = \frac{1}{\mathcal{Z}(c; W, S)} \int f(c,a) e^{W(c,a)}\, d\mu_S(a),$$

where $\mathcal{Z}(c; W, S) = \int e^{W(c,a)}\, d\mu_S(a)$ and the renormalization group map

$$\Omega(W,S)(c) = \log \frac{1}{Z_{W,S}} \int e^{W(c,a)}\, d\mu_S(a), \quad \text{where } Z_{W,S} = \int e^{W(0,a)}\, d\mu_S(a)$$

with respect to the interaction W and covariance S.

THEOREM 2.13. *Let* $\mathrm{F}, \mathrm{D} > 0$, $0 < t, v \le \frac{1}{2}$ *and* $\alpha \ge 4$. *Let* $W, V \in (\mathbf{AC})^0$. *If, for all* $\mathrm{H}, \mathrm{J} \in \bigcup_{r \ge 0} \mathcal{M}_r$

$$\left|\int b_{\mathrm{H}}{:}b_{\mathrm{J}}{:}_S \, d\mu_S(b)\right| \le \mathrm{F}^{|\mathrm{H}|+|\mathrm{J}|} \qquad \left|\int b_{\mathrm{H}}{:}b_{\mathrm{J}}{:}_T \, d\mu_T(b)\right| \le (\sqrt{t}\mathrm{F})^{|\mathrm{H}|+|\mathrm{J}|}$$
$$\|S\| \le \mathrm{F}^2\mathrm{D} \qquad \|T\| \le t\mathrm{F}^2\mathrm{D}$$
$$\mathrm{D}\|W\|_{(\alpha+2)\mathrm{F}} \le \frac{1}{6} \qquad \mathrm{D}\|V\|_{(\alpha+2)\mathrm{F}} \le \frac{v}{6}$$

then

$$|||\mathcal{S}(f; W+V, S+T) - \mathcal{S}(f; W, S)|||_{\alpha\mathrm{F}} \le 8(t+v)|||f|||_{\alpha\mathrm{F}}$$
$$\|\Omega(W+V, S+T) - \Omega(W, S)\|_{\alpha\mathrm{F}} \le \frac{3}{\mathrm{D}}(t+v).$$

PROOF. First observe that, for all $|z| \le 1/t$

$$\left|\int b_{\mathrm{H}}{:}b_{\mathrm{J}}{:}_{zT} \, d\mu_{zT}(b)\right| \le \mathrm{F}^{|\mathrm{H}|+|\mathrm{J}|} \qquad \text{by Problem 2.3(a)}$$
$$\left|\int b_{\mathrm{H}}{:}b_{\mathrm{J}}{:}_{S+zT} \, d\mu_{S+zT}(b)\right| \le (2F)^{|\mathrm{H}|+|\mathrm{J}|} \qquad \text{by Problem 2.3(b)}$$
$$\|S + zT\| \le 2\mathrm{F}^2\mathrm{D} \le (2\mathrm{F})^2\mathrm{D}.$$

Also, for all $|z'| \le 1/v$,

$$\mathrm{D}\|W + z'V\|_{(\alpha+2)\mathrm{F}} \le \frac{1}{3}$$

Hence, by Corollary 2.7, with F replaced by 2F, α replaced by $\alpha/2$, W replaced by $W + z'V$ and S replaced by $S + zT$

$$|||S(f; W+z'V, S+zT)|||_{\alpha\mathrm{F}} \le \frac{\alpha}{\alpha-2}|||f|||_{\alpha\mathrm{F}} \tag{2.2}$$
$$\|\Omega(W+z'V, S+zT)\|_{\alpha\mathrm{F}} \le \frac{\alpha}{\alpha-2}\|W+z'V\|_{\alpha\mathrm{F}} \le \frac{\alpha}{\alpha-2}\frac{1}{3\mathrm{D}}$$

for all $|z| \le 1/t$ and $|z'| \le 1/v$.

The linear map $\mathrm{R}(\cdot; \varepsilon W + z'V, S + zT)$ is, for each $0 \le \varepsilon \le 1$, a polynomial in z and z'. Bu Theorem 2.6, the kernel of $\mathbf{1} - \mathrm{R}$ is trivial for all $0 \le \varepsilon \le 1$. Hence, by Theorem 2.2 and (2.1), both $S(f; W + z'V, S + zT)$ and $\Omega(W + z'V, S + zT)$ are analytic functions of z and z' on $|z| \le 1/t$, $|z'| \le 1/v$.

By the Cauchy integral formula, if $f(z)$ is analytic and bounded in absolute value by M on $|z| \le r$, then

$$f'(z) = \frac{1}{2\pi\imath}\int_{|\zeta|=r} \frac{f(\zeta)}{(\zeta - z)^2} d\zeta$$

and, for all $|z| \le r/2$,

$$|f'(z)| \le \frac{1}{2\pi}\frac{M}{(r/2)^2}2\pi r = 4M\frac{1}{r}.$$

Hence, for all $|z| \le 1/(2t)$, $|z'| \le 1/(2v)$

$$\left|\!\left|\!\left| \frac{d}{dz} S(f; W + z'V, S + zT) \right|\!\right|\!\right|_{\alpha\mathrm{F}} \le 4t \frac{\alpha}{\alpha - 2} |||f|||_{\alpha\mathrm{F}}$$

$$\left|\!\left|\!\left| \frac{d}{dz'} S(f; W + z'V, S + zT) \right|\!\right|\!\right|_{\alpha\mathrm{F}} \le 4v \frac{\alpha}{\alpha - 2} |||f|||_{\alpha\mathrm{F}}$$

$$\left\| \frac{d}{dz} \Omega(W + z'V, S + zT) \right\|_{\alpha\mathrm{F}} \le 4t \frac{\alpha}{\alpha - 2} \frac{1}{3\mathrm{D}}$$

$$\left\| \frac{d}{dz'} \Omega(W + z'V, S + zT) \right\|_{\alpha\mathrm{F}} \le 4v \frac{\alpha}{\alpha - 2} \frac{1}{3\mathrm{D}}$$

By the chain rule, for all $|z| \le 1$,

$$\left|\!\left|\!\left| \frac{d}{dz} \mathcal{S}(f; W + zV, S + zT) \right|\!\right|\!\right|_{\alpha\mathrm{F}} \le 4(t + v) \frac{\alpha}{\alpha - 2} |||f|||_{\alpha\mathrm{F}}$$

$$\left\| \frac{d}{dz} \Omega(W + zV, S + zT) \right\|_{\alpha\mathrm{F}} \le 4(t + v) \frac{\alpha}{\alpha - 2} \frac{1}{3\mathrm{D}}.$$

Integrating z from 0 to 1 and

$$|||\mathcal{S}(f; W + V, S + T) - \mathcal{S}(f; W, S)|||_{\alpha\mathrm{F}} \le 4(t + v) \frac{\alpha}{\alpha - 2} |||f|||_{\alpha\mathrm{F}}$$

$$\|\Omega(W + V, S + T) - \Omega(W, S)\|_{\alpha\mathrm{F}} \le 4(t + v) \frac{\alpha}{\alpha - 2} \frac{1}{3\mathrm{D}}.$$

As $\alpha \ge 4$, $\alpha/(\alpha - 2) \le 2$ and the theorem follows. □

2.4. Sample Applications

We now apply the expansion to a few examples. In these examples, the set of Grassmann algebra generators $\{a_1, \ldots, a_n\}$ is replaced by $\{\psi_{x,\sigma}, \bar{\psi}_{x,\sigma} \mid x \in \mathbb{R}^{d+1}, \sigma \in \mathfrak{G}\}$ with $\mathfrak{G}$ being a finite set (of spin/color values). See Section 1.5. To save writing, we introduce the notation

$$\xi = (x, \sigma, b) \in \mathbb{R}^{d+1} \times \mathfrak{G} \times \{0, 1\}$$

$$\int d\xi \cdots = \sum_{b \in \{0,1\}} \sum_{\sigma \in \mathfrak{G}} \int d^{d+1}x \cdots$$

$$\psi_\xi = \begin{cases} \psi_{x,\sigma}, & \text{if } b = 0; \\ \bar{\psi}_{x,\sigma}, & \text{if } b = 1. \end{cases}$$

Now elements of the Grassmann algebra have the form

$$f(\psi) = \sum_{r=0}^{\infty} \int d\xi_1 \cdots d\xi_r \, f_r(\xi_1, \ldots, \xi_r) \psi_{\xi_1} \cdots \psi_{\xi_r}$$

the norms $\|f_r\|$ and $\|f(\psi)\|_\alpha$ become

$$\|f_r\| = \max_{1 \le i \le r} \sup_{\xi_i} \int \prod_{\substack{j=1 \\ j \ne i}}^{r} d\xi_j \, |f_r(\xi_1, \ldots, \xi_r)|$$

$$\|f(\psi)\|_\alpha = \sum_{r=0}^{\infty} \alpha^r \|f_r\|.$$

When we need a second copy of the generators, we use $\{\Psi_\xi \mid \xi \in \mathbb{R}^{d+1} \times \mathfrak{G} \times \{0,1\}\}$ (in place of $\{c_1, \ldots, c_n\}$). Then elements of the Grassmann algebra have the form

$$f(\Psi, \psi) = \sum_{l,r=0}^{\infty} \int d\xi_1' \cdots d\xi_l' d\xi_1 \cdots d\xi_r \, f_{l,r}(\xi_1', \ldots, \xi_l'; \xi_1, \ldots, \xi_r) \Psi_{\xi_1'} \cdots \Psi_{\xi_l'} \psi_{\xi_1} \cdots \psi_{\xi_r}$$

and the norms $\|f_{l,r}\|$ and $\|f(\Psi, \psi)\|_\alpha$ are

$$\|f_{l,r}\| = \max_{1 \le i \le l+r} \sup_{\xi_i} \int \prod_{\substack{j=1 \\ j \ne i}}^{l+r} d\xi_j \, |f_{l,r}(\xi_1, \ldots, \xi_l; \xi_{l+1}, \ldots, \xi_{l+r})|$$

$$\|f(\Psi, \psi)\|_\alpha = \sum_{l,r=0}^{\infty} \alpha^{l+r} \|f_{l,r}\|.$$

All of our Grassmann Gaussian Integrals will obey

$$\int \psi_{x,\sigma} \psi_{x',\sigma'} \, d\mu_S(\psi) = 0, \qquad \int \bar\psi_{x,\sigma} \bar\psi_{x',\sigma'} \, d\mu_S(\psi) = 0,$$

$$\int \psi_{x,\sigma} \bar\psi_{x',\sigma'} \, d\mu_S(\psi) = -\int \bar\psi_{x',\sigma'} \psi_{x,\sigma} \, d\mu_S(\psi).$$

Hence, if $\xi = (x, \sigma, b)$, $\xi' = (x', \sigma', b')$

$$S(\xi, \xi') = \begin{cases} 0, & \text{if } b = b' = 0; \\ C_{\sigma,\sigma'}(x, x'), & \text{if } b = 0,\ b' = 1; \\ -C_{\sigma',\sigma}(x', x), & \text{if } b = 1,\ b' = 0; \\ 0, & \text{if } b = b' = 1, \end{cases}$$

where

$$C_{\sigma,\sigma'}(x, x') = \int \psi_{x,\sigma} \bar\psi_{x,\sigma'} \, d\mu_S(\psi).$$

That our Grassmann algebra is no longer finite-dimensional is, in itself, *not* a big deal. The Grassmann algebras are not the ultimate objects of interest. The ultimate objects of interest are various expectation values. These expectation values are complex numbers that we have chosen to express as the values of Grassmann Gaussian integrals. See (1.3). If the covariances of interest were to satisfy the hypotheses (HG) and (HS), we would be able to easily express the expectation values as limits of integrals over finite-dimensional Grassmann algebras using Corollary 2.7 and Theorem 2.13.

The real difficulty is that for many, perhaps most, models of interest, the covariances (also called propagators) do not satisfy (HG) and (HS). So, as explained in Section 1.5, we express the covariance as a sum of terms, each of which does satisfy the hypotheses. These terms, called single scale covariances, will, in each example, be constructed by substituting a partition of unity of $\mathbb{R}^{d+1}$ (momentum space) into the full covariance. The partition of unity will be constructed using a fixed "scale parameter" $M > 1$ and a function $\nu \in C_0^\infty([M^{-2}, M^2])$ that takes values in $[0, 1]$, is identically 1 on $[M^{-1/2}, M^{1/2}]$ and obeys

$$\sum_{j=0}^{\infty} \nu(M^{2j} x) = 1 \tag{2.3}$$

for $0 < x < 1$.

PROBLEM 2.6. Let $M > 1$. Construct a function $\nu \in C_0^\infty([M^{-2}, M^2])$ that takes values in $[0,1]$, is identically 1 on $[M^{-1/2}, M^{1/2}]$ and obeys

$$\sum_{j=0}^{\infty} \nu(M^{2j}x) = 1$$

for $0 > x < 1$.

EXAMPLE (Gross-Neveu$_2$). The propagator for the Gross-Neveu model in two space-time dimensions has

$$C_{\sigma,\sigma'}(x,x') = \int \psi_{x,\sigma}\bar{\psi}_{x,\sigma'}\, d\mu_S(\psi) = \int \frac{d^2p}{(2\pi)^2} e^{ip\cdot(x'-x)} \frac{\not{p}_{\sigma,\sigma'} + m\delta_{\sigma,\sigma'}}{p^2+m^2}$$

where

$$\not{p} = \begin{pmatrix} ip_0 & p_1 \\ -p_1 & -ip_0 \end{pmatrix}$$

is a 2×2 matrix whose rows and columns are indexed by $\sigma \in \{\uparrow, \downarrow\}$. This propagator does not satisfy Hypothesis (HG) for any finite F. If it did satisfy (HG) for some finite F, $C_{\sigma,\sigma'}(x,x') = \int \psi_{x,\sigma}\bar{\psi}_{x',\sigma'}\, d\mu_S(\psi)$ would be bounded by F^2 for all x and x'. This is not the case—it blows up as $x' - x \to 0$.

Set

$$\nu_j(p) = \begin{cases} \nu\left(\frac{M^{2j}}{p^2}\right), & \text{if } j > 0; \\ \nu\left(\frac{M^{2j}}{p^2}\right), & \text{if } j = 0,\ |p| \geq 1. \\ 1, & \text{if } j = 0,\ |p| < 1. \end{cases}$$

Then

$$S(x,y) = \sum_{j=0}^{\infty} S^{j)}(x,y)$$

with

$$S^{(j)}(\xi,\xi') = \begin{cases} 0, & \text{if } b = b' = 0; \\ C^{(j)}_{\sigma,\sigma'}(x,x'), & \text{if } b = 0,\ b' = 1; \\ -C^{(j)}_{\sigma',\sigma}(x',x), & \text{if } b = 1,\ b' = 0; \\ 0, & \text{if } b = b' = 1 \end{cases}$$

and

$$C^{(j)}(x,x') = \int \frac{d^2p}{(2\pi)^2} e^{ip\cdot(x'-x)} \frac{\not{p}+m}{p^2+m^2}\nu_j(p). \tag{2.4}$$

We now check that, for each $0 \leq j < \infty$, $S^{(j)}$ does satisfy Hypotheses (HG) and (HS). The integrand of $C^{(j)}$ is supported on $M^{j-1} \leq |p| \leq M^{j+1}$ for $j > 0$ and $|p| \leq M$ for $j = 0$. This is a region of volume at most $\pi(M^{j+1})^2 \leq \text{const}\, M^{2j}$. By Corollary 1.35 and Problem 2.7, below, the value of F for this propagator is bounded by

$$\mathrm{F}_j = \left(2\int \left\| \frac{\not{p}+m}{p^2+m^2} \right\| \nu_j(p) \frac{d^2p}{(2\pi)^2}\right)^{1/2} \leq C_{\mathrm{F}}\left(\frac{1}{M^j}M^{2j}\right)^{1/2} = C_{\mathrm{F}} M^{j/2}$$

for some constant C_{F}. Here $\|(\not{p}+m)/(p^2+m^2)\|$ is the matrix norm of $(\not{p}+m)/(p^2+m^2)$.

PROBLEM 2.7. Prove that

$$\left\| \frac{\not p + m}{p^2 + m^2} \right\| = \frac{1}{\sqrt{p^2 + m^2}}.$$

By the following lemma, the value of D for this propagator is bounded by

$$\mathrm{D}_j = \frac{1}{M^{2j}}.$$

We have increased the value of C_{F} in F_j in order to avoid having a const in D_j.

LEMMA 2.14.

$$\sup_{x,\sigma} \sum_{\sigma'} \int d^2y \, |C^{(j)}_{\sigma,\sigma'}(x, y)| \le \text{const} \, \frac{1}{M^j}.$$

PROOF. When $j = 0$, the lemma reduces to $\sup_{x,\sigma} \sum_{\sigma'} \int d^2y \, |C^{(0)}_{\sigma,\sigma'}(x, y)| < \infty$. This is the case, because every matrix element of $[(\not p + m)/(p^2 + m^2)]\nu_j(p)$ is C^∞_0, so that $C^{(0)}_{\sigma,\sigma'}(x, y)$ is a C^∞ rapidly decaying function of $x - y$. Hence it suffices to consider $j \ge 1$. The integrand of (2.4) is supported on a region of volume at most $\pi(M^{j+1})^2 \le \text{const}\, M^{2j}$ and has every matrix element bounded by

$$\frac{M^{j+1} + m}{M^{2j-2} + m^2} \le \text{const} \, \frac{1}{M^j}$$

since $M^{j-1} \le |p| \le M^{j+1}$ on the support of $\nu_j(p)$. Hence

$$\begin{aligned}
(2.5) \qquad \sup_{\substack{x,y \\ \sigma,\sigma'}} |C^{(j)}_{\sigma,\sigma'}(x, y)| &\le \int \sup_{\sigma,\sigma'} \left| \left(\frac{\not p + m}{p^2 + m^2} \right)_{\sigma,\sigma'} \right| \nu_j(p) \frac{d^2p}{(2\pi)^2} \\
&\le \text{const} \, \frac{1}{M^j} \int \nu_j(p)\, d^2p \le \text{const} \, \frac{1}{M^j} M^{2j} \le \text{const}\, M^j.
\end{aligned}$$

To show that $C^{(j)}_{\sigma,\sigma'}(x, y)$ decays sufficiently quickly in $x - y$, we play the usual integration by parts game.

$$\begin{aligned}
(y - x)^4 C^{(j)}_{\sigma,\sigma'}(x, y) &= \int \frac{d^2p}{(2\pi)^2} \frac{\not p + m}{p^2 + m^2} \nu_j(p) \left(\frac{\partial^2}{\partial p_1^2} + \frac{\partial^2}{\partial p_2^2} \right)^2 e^{ip\cdot(y-x)} \\
&= \int \frac{d^2p}{(2\pi)^2} e^{ip\cdot(y-x)} \left(\frac{\partial^2}{\partial p_1^2} + \frac{\partial^2}{\partial p_2^2} \right)^2 \left(\frac{\not p + m}{p^2 + m^2} \nu_j(p) \right).
\end{aligned}$$

Each matrix element of $(\not p + m)/(p^2 + m^2)$ is a ratio $P(p)/Q(p)$ of two polynomials in $p = (p_0, p_1)$. For any such rational function

$$\frac{\partial}{\partial p_i} \frac{P(p)}{Q(p)} = \frac{\partial P(p)/\partial p_i Q(p) - P(p) \partial Q(p)/\partial p_i(p)}{Q(p)^2}.$$

The difference between degrees of the numerator and denominator of $\partial[P(p)/Q(p)]/\partial p_i$ obeys

$$\deg\left(\frac{\partial P}{\partial p_i} Q - P \frac{\partial Q}{\partial p_i} \right) - \deg Q^2 \le \deg P + \deg Q - 1 - 2 \deg Q = \deg P - \deg Q - 1.$$

Hence

$$\frac{\partial^{\alpha_0}}{\partial p_0^{\alpha_0}} \frac{\partial^{\alpha_1}}{\partial p_1^{\alpha_1}} \left(\frac{\not p + m}{p^2 + m^2} \right)_{\sigma,\sigma'} = \frac{P^{\alpha_0,\alpha_1)}_{\sigma,\sigma'}(p)}{(p^2 + m^2)^{1+\alpha_0+\alpha_1}}$$

with $P^{(\alpha_0,\alpha_1)}_{\sigma,\sigma'}(p)$ a polynomial of degree at most $\deg\big((p^2+m^2)^{1+\alpha_0+\alpha_1}\big)-1-\alpha_0-\alpha_1 = 1+\alpha_0+\alpha_1$. In words, each $\partial/\partial p_i$ acting on $(\not p+m)/(p^2+m^2)$ increases the difference between the degree of the denominator and the degree of the numerator by one. So does each $\partial/\partial p_i$ acting on $\nu_j(p)$, provided you count both M^j and p as having degree one. For example,

$$\frac{\partial}{\partial p_0}\nu\Big(\frac{M^{2j}}{p^2}\Big) = -\frac{2p_0}{p^4}M^{2j}\nu'\Big(\frac{M^{2j}}{p^2}\Big)$$
$$\frac{\partial^2}{\partial p_0^2}\nu\Big(\frac{M^{2j}}{p^2}\Big) = \Big[-\frac{2}{p^4}+\frac{8p_0^2}{p^6}\Big]M^{2j}\nu'\Big(\frac{M^{2j}}{p^2}\Big)+\frac{4p_0^2}{p^8}M^{4j}\nu'\Big(\frac{M^{2j}}{p^2}\Big).$$

In general,

$$\Big(\frac{\partial^2}{\partial p_1^2}+\frac{\partial^2}{\partial p_2^2}\Big)^2\Big(\frac{\not p_{\sigma,\sigma'}+m}{p^2+m^2}\nu_j(p)\Big) = \sum_{\substack{n,\ell\in\mathbb{N}\\ 1\le n+\ell\le 4}} \frac{P_{\sigma,\sigma',n,\ell}(p)}{(p^2+m^2)^{1+n}}\frac{M^{2\ell j}Q_{n,\ell}(p)}{p^{4\ell+2(4-n-\ell)}}\nu^{(\ell)}\Big(\frac{M^{2j}}{p^2}\Big).$$

Here n is the number of derivatives that acted on $(\not p_{\sigma,\sigma'}+m)/(p^2+m^2)$ and ℓ is the number of derivatives that acted on ν. The remaining $4-n-\ell$ derivatives acted on the factors arising from the argument of ν upon application of the chain rule. The polynomials $P_{\sigma,\sigma',n,\ell}(p)$ and $Q_{n,\ell}(p)$ have degrees at most $1+n$ and $\ell+(4-n-\ell)$, respectively. All together, when you count both M^j and p as having degree one, the degree of the denominator $(p^2+m^2)^{1+n}p^{4\ell+2(4-n-\ell)}$ namely $2(1+n)+4\ell+2(4-n-\ell) = 10+2\ell$, is at least five times larger than the degree of the numerator $P_{\sigma,\sigma',n,\ell}(p)M^{2\ell j}Q_{n,\ell}(p)$, which is at most $1+n+2\ell+\ell+(4-n-\ell) = 5+2\ell$. Recalling that $|p|$ is bounded above and below by const M^j (of course with different constants),

$$\Big|\frac{P_{\sigma,\sigma',n,\ell}(p)}{(p^2+m^2)^{1+n}}\frac{M^{2\ell j}Q_{n,\ell}(p)}{p^{4\ell+2(4-n-\ell)}}\nu^{(\ell)}\Big(\frac{M^{2j}}{p^2}\Big)\Big| \le \text{const}\,\frac{M^{(1+n)j}M^{2\ell j}M^{(4-n)j}}{M^{2j(1+n)}M^{j(8-2n+2\ell)}}$$
$$= \text{const}\,\frac{1}{M^{5j}}$$

and

$$\Big|\Big(\frac{\partial^2}{\partial p_1^2}+\frac{\partial^2}{\partial p_2^2}\Big)^2\Big(\frac{\not p+m}{p^2+m^2}\nu_j(p)\Big)\Big| \le \text{const}\,\frac{1}{M^{5j}}$$

on the support of the integrand. The support of

$$\Big(\frac{\partial^2}{\partial p_1^2}+\frac{\partial^2}{\partial p_2^2}\Big)^2\Big(\frac{\not p+m}{p^2+m^2}\nu_j(p)\Big)$$

is contained in the support of $\nu_j(p)$. So the integrand is still supported in a region of volume const M^{2j} and

$$\sup_{\substack{x,y\\ \sigma,\sigma'}}|M^{4j}(y-x)^4C^{(j)}_{\sigma,\sigma'}(x,y)| \le \text{const}\,M^{4j}\frac{1}{M^{5j}}M^{2j} \le \text{const}\,M^j. \tag{2.6}$$

Multiplying the $\frac{1}{4}$ power of (2.5) by the $\frac{3}{4}$ power of (2.6) gives

$$\sup_{\substack{x,y\\ \sigma,\sigma'}}\big|M^{3j}|y-x|^3C^{(j)}_{\sigma,\sigma'}(x,y)\big| \le \text{const}\,M^j. \tag{2.7}$$

Adding (2.5) to (2.7) gives

$$\sup_{\substack{x,y\\ \sigma,\sigma'}}\big|[1+M^{3j}|y-x|^3]C^{(j)}_{\sigma,\sigma'}(x,y)\big|\le \text{const}\, M^j.$$

Dividing across

$$|C^{(j)}_{\sigma,\sigma'}(x,y)|\le \text{const}\,\frac{M^j}{1+M^{3j}|x-y|^3}.$$

Integrating

$$\begin{aligned}\int d^2y\,|C^{(j)}_{\sigma,\sigma'}(x,y)| &\le \int d^2y\,\text{const}\,\frac{M^j}{1+M^{3j}|x-y|^3}\\ &= \text{const}\,\frac{1}{M^j}\int d^2z\frac{1}{1+|z|^3}\\ &\le \text{const}\,\frac{1}{M^j}.\end{aligned}$$

We made the change of variables $z=M^j(y-x)$. □

To apply Corollary 2.7 to this model, we fix some $\alpha\ge 2$ and define the norm $\|W\|_j$ of $W(\psi)=\sum_{r>0}\int d\xi_1\cdots d\xi_r\, w_r(\xi_1,\ldots,\xi_r)\psi_{\xi_1}\cdots\psi_{\xi_r}$ to be

$$\|W\|_j=\mathrm{D}_j\|W\|_{\alpha\mathrm{F}_j}=\sum_r(\alpha C_{\mathrm{F}})^r M^{j(r-4)/2}\|w_r\|.$$

Let $J>0$ be a cutoff parameter (meaning that, in the end, the model is defined by taking the limit $J\to\infty$) and define, as in Section 1.5,

$$\mathcal{G}_J(\Psi)=\log\frac{1}{Z_J}\int e^{W(\Psi+\psi)}\,d\mu_{S^{(\le J)}}(\psi),\quad \text{where } Z_J=\int e^{W(\psi)}\,d\mu_{S^{(\le J)}}(\psi)$$

and

$$\Omega_j(W)(\Psi)=\log\frac{1}{Z_{W,S^{(j)}}}\int e^{W(\Psi+\psi)}\,d\mu_{S^{(j)}}(\psi),$$

where $Z_{W,S^{(j)}}=\int e^{W(\psi)}\,d\mu_{S^{(j)}}(\psi)$. Then, by Problem 1.16

$$\mathcal{G}_J=\Omega_{S^{(1)}}\circ\Omega_{S^{(2)}}\circ\cdots\circ\Omega_{S^{(J)}}(W).$$

Set

$$W_j=\Omega_{S^{(j)}}\circ\Omega_{S^{(j+1)}}\circ\cdots\circ\Omega_{S^{(J)}}(W).$$

Suppose that we have shown that $\|W_j\|_j\le\frac{1}{3}$. To integrate out scale $j-1$ we use

THEOREM 2.15 GN. *Suppose $\alpha\ge 2$ and $M\ge\big(\alpha/(\alpha-1)\big)(2(\alpha+1)/\alpha)^6$. If $\|W\|_j\le\frac{1}{3}$ and w_r vanishes for $r\le 4$, then $\|\Omega_{j-1}(W)\|_{j-1}\le\|W\|_j$.*

PROOF. We first have to relate $\|W(\Psi+\psi)\|_\alpha$ to $\|W(\psi)\|_\alpha$, because we wish to apply Corollary 2.7 with $W(c,a)$ replaced by $W(\Psi+\psi)$. To do so, we temporarily revert to the old notation with c and a generators, rather than Ψ and ψ generators. Observe that

$$\begin{aligned}W(c+a)&=\sum_m\sum_{\mathrm{I}\in\mathcal{M}_m} w_m(\mathrm{I})(c+a)_{\mathrm{I}}\\ &=\sum_m\sum_{\mathrm{I}\in\mathcal{M}_m}\sum_{\mathrm{J}\subset\mathrm{I}} w_m\big(\mathrm{J}(\mathrm{I}\setminus\mathrm{J})\big)c_{\mathrm{J}}a_{\mathrm{I}\setminus\mathrm{J}}\\ &=\sum_{l,r}\sum_{\mathrm{J}\in\mathcal{M}_l}\sum_{\mathrm{K}\in\mathcal{M}_r}\binom{l+r}{l}w_{l+r}(\mathrm{JK})c_{\mathrm{J}}a_{\mathrm{K}}.\end{aligned}$$

In passing from the first line to the second line, we used that $w_m(\mathrm{I})$ is antisymmetric under permutation of its arguments. In passing from the second line to the third line, we renamed $\mathrm{I}\setminus\mathrm{J}=\mathrm{K}$. The $\binom{l+r}{l}$ arises because, given two ordered sets J, K, there are $\binom{|\mathrm{J}|+|\mathrm{K}|}{|\mathrm{J}|}$ ordered sets I with $\mathrm{J}\subset\mathrm{I}$, $\mathrm{K}=\mathrm{I}\setminus\mathrm{J}$. Hence

$$\|W(c+a)\|_\alpha=\sum_{l,r}\alpha^{l+r}\binom{l+r}{l}\|w_{l+r}\|=\sum_m\alpha^m 2^m\|w_m\|=\|W(a)\|_{2\alpha}.$$

Similarly,

$$\|W(\Psi+\psi)\|_\alpha=\|W(\psi)\|_{2\alpha}.$$

To apply Corollary 2.7 at scale $j-1$, we need

$$\mathrm{D}_{j-1}\|W(\Psi+\psi)\|_{(\alpha+1)F_{j-1}}=\mathrm{D}_{j-1}\|W(\psi)\|_{2(\alpha+1)F_{j-1}}\le\frac{1}{3}.$$

But

$$\begin{aligned}\mathrm{D}_{j-1}\|W\|_{2(\alpha+1)F_{j-1}}&=\sum_r\left(2(\alpha+1)C_{\mathrm{F}}\right)^r M^{(j-1)(r-4)/2}\|w_r\|\\&=\sum_{r\ge6}\left(2\frac{\alpha+1}{\alpha}M^{-(1/2-2/r)}\right)^r(\alpha C_{\mathrm{F}})^r M^{j(r-4)/2}\|w_r\|\\&\le\sum_{r\ge6}\left(2\frac{\alpha+1}{\alpha}M^{-1/6}\right)^r(\alpha C_{\mathrm{F}})^r M^{j(r-4)/2}\|w_r\|\\&\le\left(2\frac{\alpha+1}{\alpha}\right)^6\frac{1}{M}\|W\|_j\le\frac{1}{3}\end{aligned}$$

as $M>(2(\alpha+1)/\alpha)^6$ and $\|W\|_j\le\frac{1}{3}$. By Corollary 2.7,

$$\begin{aligned}\|\Omega_{j-1}(W)\|_{j-1}=\mathrm{D}_{j-1}\|\Omega_{j-1}(W)\|_{\alpha\mathrm{F}_{j-1}}&\le\frac{\alpha}{\alpha-1}\mathrm{D}_{j-1}\|W(\Psi+\psi)\|_{\alpha\mathrm{F}_{j-1}}\\&\le\frac{\alpha}{\alpha-1}\left(2\frac{\alpha+1}{\alpha}\right)^6\frac{1}{M}\|W\|_j\le\|W\|_j.\end{aligned}$$

□

Theorem 2.15 GN is just one ingredient used in the construction of the Gross-Neveu$_2$ model. It basically reduces the problem to the study of the projection

$$PW(\psi)=\sum_{r=2,4}\int d\xi_1\cdots d\xi_r\,w_r(\xi_1,\ldots,\xi_r)\psi_{\xi_1}\cdots\psi_{\xi_r}$$

of W onto the part of the Grassmann algebra of degree at most four. A souped up, "renormalized," version of Theorem 2.15 GN can be used to reduce the problem to the study of the projection $P'W$ of W onto a three-dimensional subspace of the range of P.

Example (Naive Many-fermion$_2$). The propagator, or covariance, for many-fermion models is

$$C_{\sigma,\sigma'}(x,x')=\delta_{\sigma,\sigma'}\int\frac{d^3k}{(2\pi)^3}e^{ik\cdot(x'-x)}\frac{1}{ik_0-e(\mathbf{k})}$$

where $k=(k_0,\mathbf{k})$ and $e(\mathbf{k})$ is the one particle dispersion relation (a generalization of $k^2/2m$) minus the chemical potential (which controls the density of the gas). The subscript on many-fermion$_2$ signifies that the number of space dimensions is two (i.e., $\mathbf{k}\in\mathbb{R}^2$, $k\in\mathbb{R}^3$). For pedagogical reasons, I am not using the standard many-body Fourier transform conventions. We assume that $e(\mathbf{k})$ is a reasonable smooth function (for example, C^4) that has a nonempty, compact, strictly convex zero set, called the Fermi curve and denoted $\mathcal{F}$. We further assume that $\nabla e(\mathbf{k})$ does not vanish for $\mathbf{k} \in \mathcal{F}$, so that $\mathcal{F}$ is itself a reasonably smooth curve. At low temperatures only those momenta with $k_0 \approx 0$ and $\mathbf{k}$ near $\mathcal{F}$ are important, so we replace the above propagator with

$$C_{\sigma,\sigma'}(x,x') = \delta_{\sigma,\sigma'} \int \frac{d^3k}{(2\pi)^3} e^{ik\cdot(x'-x)} \frac{U(k)}{ik_0 - e(\mathbf{k})}.$$

The precise ultraviolet cutoff, $U(k)$, shall be chosen shortly. It is a C_0^∞ function which take values in $[0,1]$, is identically 1 for $k_0^2 + e(\mathbf{k})^2 \le 1$ and vanishes for $k_0^2 + e(\mathbf{k})^2$ larger than some constant. This covariance does not satisfy Hypotheses (HS) for any finite D. If it did, $C_{\sigma,\sigma'}(0,x')$ would be L^1 in x' and consequently the Fourier transform $U(k)/\big(ik_0 - e(\mathbf{k})\big)$ would be uniformly bounded. But $U(k)/\big(ik_0 - e(\mathbf{k})\big)$ blows up at $k_0 = 0$, $e(\mathbf{k}) = 0$. So we write the covariance as sum of infinitely many "single scale" covariances, each of which does satisfy (HG) and (HS). This decomposition is implemented through a partition of unity of the set of all k's with $k_0^2 + e(\mathbf{k})^2 \le 1$.

We slice momentum space into shells around the Fermi curve. The jth shell is defined to be the support of

$$\nu^{(j)}(k) = \nu\big(M^{2j}(k_0^2 + e(\mathbf{k})^2)\big)$$

where ν is the function of (2.3). By construction, $\nu(x)$ vanishes unless $1/M^2 \le x \le M^2$, so that the jth shell is a subset of

$$\left\{ k \;\middle|\; \frac{1}{M^{j+1}} \le |ik_0 - e(\mathbf{k})| \le \frac{1}{M^{j-1}} \right\}$$

As the scale parameter $M > 1$, the shells near the Fermi curve have j near $+\infty$. Setting

$$C^{(j)}_{\sigma,\sigma'}(x,x') = \delta_{\sigma,\sigma'} \int \frac{d^3k}{(2\pi)^3} e^{ik\cdot(x'-x)} \frac{\nu^{(j)}(k)}{ik_0 - e(\mathbf{k})}$$

and $U(k) = \sum_{j=0}^\infty \nu^{(j)}(k)$ we have

$$C_{\sigma,\sigma'}(x,x') = \sum_{j=0}^\infty C^{(j)}_{\sigma,\sigma'}(x,x').$$

The integrand of the propagator $C^{(j)}$ is supported on a region of volume at most const M^{-2j} ($|k_0| \le 1/M^{j-1}$ and, as $|e(\mathbf{k})| \le 1/M^{j-1}$ and ∇e is nonvanishing on $\mathcal{F}$, $\mathbf{k}$ must remain within a distance const M^{-j} of $\mathcal{F}$) and is bounded by M^{j+1}. By Corollary 1.35, the value of F for this propagator is bounded by

$$\text{(2.8)} \qquad \mathrm{F}_j = \left(2 \int \frac{\nu^{(j)}(k)}{|ik_0 - e(\mathbf{k})|} \frac{d^3k}{(2\pi)^3} \right)^{1/2} \le C_{\mathrm{F}} \left(M^j \frac{1}{M^{2j}} \right)^{1/2} = C_{\mathrm{F}} \frac{1}{M^{j/2}}$$

for some constant C_{F}. Also

$$\sup_{\substack{x,y \\ \sigma,\sigma'}} |C^{(j)}_{\sigma,\sigma'}(x,y)| \le \int \frac{\nu^{(j)}(k)}{|ik_0 - e(\mathbf{k})|} \frac{d^3k}{(2\pi)^3} \le \text{const}\, \frac{1}{M^j}.$$

Each derivative $\partial/\partial k_i$ acting on $\nu^{(j)}(k)/ik_0 - e(\mathbf{k})$ increases the supremum of its magnitude by a factor of order M^j. So the naive argument of Lemma 2.14 gives

$$|C^{(j)}_{\sigma,\sigma'}(x,y)| \le \text{const}\,\frac{1/M^j}{[1+M^{-j}|x-y|]^4} \implies \sup_{x,\sigma}\sum_{\sigma'}\int d^3y\,|C^{(j)}_{\sigma,\sigma'}(x,y)| \le \text{const}\,M^{2j}.$$

There is not much point going through this bound in greater detail, because Corollary C.3 gives a better bound. In Appendix C, we express, for any $\mathfrak{l}_j \in [M^{-j}, M^{-j/2}]$, $C^{(j)}_{\sigma,\sigma'}(x,y)$ as a sum of at most const$/\mathfrak{l}_j$ terms, each of which is bounded in Corollary C.3. Applying that bound, with $\mathfrak{l} = 1/M^{j/2}$, yields the better bound

$$\sup_{x,\sigma}\sum_{\sigma'}\int d^3y\,|C^{(j)}_{\sigma,\sigma'}(x,y)| \le \text{const}\,\frac{1}{\mathfrak{l}_j}M^j \le \text{const}\,M^{3j/2}. \tag{2.9}$$

So the value of D for this propagator is bounded by

$$\mathrm{D}_j = M^{5j/2}.$$

This time we define the norm

$$\|W\|_j = \mathrm{D}_j\|W\|_{\alpha \mathrm{F}_j} = \sum_r (\alpha C_{\mathrm{F}})^r M^{-j(r-5)/2}\|w_r\|$$

Again, let $J > 0$ be a cutoff parameter and define, as in Section 1.5,

$$\mathcal{G}_J(c) = \log\frac{1}{Z_J}\int e^{W(\Psi+\psi)}\,d\mu_{S^{(\le J)}}(a), \quad \text{where } Z_J = \int e^{W(\psi)}\,d\mu_{S^{(\le J)}}(a)$$

and

$$\Omega_j(W)(c) = \log\frac{1}{Z_{W,S^{(j)}}}\int e^{W(\Psi+\psi)}\,d\mu_{S^{(\le j)}}(a),$$

where $Z_{W,S^{(j)}} = \int e^{W(\psi)}\,d\mu_{S^{(j)}}(a)$. Then, by Problem 1.16,

$$\mathcal{G}_J = \Omega_{S^{(J)}} \circ \cdots \circ \Omega_{S^{(1)}} \circ \Omega_{S^{(0)}}(W).$$

Also call $\mathcal{G}_j = W_j$. If we have integrated out all scales from the ultraviolet cutoff, which in this (infrared) problem is fixed at scale 0, to j and we have ended up with some interaction that obeys $\|W\|_j \le \frac{1}{3}$, then we integrate out scale $j+1$ using the following analog of Theorem 2.15 GN.

THEOREM 2.15 MB$_1$. *Suppose $\alpha \ge 2$ and $M \ge \big(2\alpha/(\alpha-1)\big)^2\big((\alpha+1)/\alpha\big)^{12}$. If $\|W\|_j \ge \frac{1}{3}$ and w_r vanishes for $r < 6$, then $\|\Omega_{j+1}(W)\|_{j+1} \le \|W\|_j$.*

PROOF. To apply Corollary 2.7 at scale $j+1$, we need

$$\mathrm{D}_{j+1}\|W(\Psi+\psi)\|_{(\alpha+1)\mathrm{F}_{j+1}} = \mathrm{D}_{j+1}\|W\|_{2(\alpha+1)\mathrm{F}_{j+1}} \le \frac{1}{3}.$$

But

$$\begin{aligned}
\mathrm{D}_{j+1}\|W\|_{2(\alpha+1)\mathrm{F}_{j+1}} &= \sum_r \big(2(\alpha+1)C_{\mathrm{F}}\big)^r M^{-(j+1)(r-5)/2}\|w_r\| \\
&= \sum_{r\ge 6}\Big(2\frac{\alpha+1}{\alpha}M^{-(1-5/r)/2}\Big)^r (\alpha C_{\mathrm{F}})^r M^{-j(r-5)/2}\|w_r\| \\
&\le \sum_{r\ge 6}\Big(2\frac{\alpha+1}{\alpha}M^{-1/12}\Big)^r (\alpha C_{\mathrm{F}})^r M^{-j(r-5)/2}\|w_r\| \\
&\le \Big(2\frac{\alpha+1}{\alpha}\Big)^6 \frac{1}{M^{1/2}}\|W\|_j \le \|W\|_j \le \frac{1}{3}.
\end{aligned}$$

By Corollary 2.7

$$\begin{aligned}\|\Omega_{j+1}(W)\|_{j+1} &= \mathrm{D}_{j+1}\|\Omega_{j+1}(W)\|_{\alpha\mathrm{F}_{j+1}}\\ &\le \frac{\alpha}{\alpha-1}\mathrm{D}_{j+1}\|W(\Psi+\psi)\|_{\alpha\mathrm{F}_{j+1}}\\ &\le \frac{\alpha}{\alpha-1}\left(2\frac{\alpha+1}{\alpha}\right)^6\frac{1}{M^{1/2}}\|W\|_j \le \|W\|_j. \qquad \square\end{aligned}$$

It looks, in Theorem 2.15 MB_1, like five-legged vertices w_5 are marginal and all vertices w_r with $r<5$ have to be renormalized. Of course, by evenness, there are no five-legged vertices so only vertices w_r with $r=2,4$ have to be renormalized. But it still looks, contrary to the behavior of perturbation theory [**FST**], like four-legged vertices are worse than marginal. Fortunately, this is not the case. Our bounds can be tightened still further.

In the bounds (2.8) and (2.9) the momentum k runs over a shell around the Fermi curve. Effectively, the estimates we have used to count powers of M^j assume that all momenta entering an r-legged vertex run independently over the shell. Thus the estimates fail to take into account conservation of momentum. As a simple illustration of this, observe that for the two-legged diagram $B(x,y)=\int d^3z\, C^{(j)}_{\sigma,\sigma}(x,z)C(j)_{\sigma,\sigma}(z,y)$, (2.9) yields the bound

$$\begin{aligned}\sup_x\int d^3y\,|B(x,y)| &\le \sup_x\int d^3z|C^{(j)}_{\sigma,\sigma}(x,z)|\int d^3y\,|C^{(j)}_{\sigma,\sigma}(z,y)|\\ &\le \mathrm{const}\,M^{3j/2}M^{3j/2} = \mathrm{const}\,M^{3j}.\end{aligned}$$

But $B(x,y)$ is the Fourier transform of

$$W(k)=\frac{\nu^{(j)}(k)^2}{[ik_0-e(\mathbf{k})]^2}=C^{(j)}(k)C^{(j)}(p)|_{p=k}.$$

Conservation of momentum forces the momenta in the two lines to be the same. Plugging this $W(k)$ and $\mathfrak{l}_j=1/M^{j/2}$ into Corollary C.2 yields

$$\sup_x\int d^3y\,|B(x,y)|\le \mathrm{const}\,\frac{1}{\mathfrak{l}_j}M^{2j}\le \mathrm{const}\,M^{5j/2}.$$

We exploit conservation of momentum by partitioning the Fermi curve into "sectors."

Example (Many-fermion$_2$—with sectorization). We start by describing precisely what sectors are, as subsets of momentum space. Let, for $k=(k_0,\mathbf{k})$, $\mathbf{k}'(k)$ be any reasonable "projection" of $\mathbf{k}$ onto the Fermi curve

$$\mathcal{F}=\{\mathbf{k}\in\mathbb{R}^2 \mid e(\mathbf{k})=0\}.$$

In the event that $\mathcal{F}$ is a circle of radius $k_{\mathcal{F}}$ centered on the origin, it is natural to choose $\mathbf{k}'(k)=(k_{\mathcal{F}}/|\mathbf{k}|)\mathbf{k}$. For general $\mathcal{F}$, one can always construct, in a tubular neighborhood of $\mathcal{F}$, a C^∞ vector field that is transverse to $\mathcal{F}$, and then define $\mathbf{k}'(k)$ to be unique point of $\mathcal{F}$ that is on the same integral curve of the vector field as $\mathbf{k}$ is.

Let $j>0$ and set

$$\nu^{(\ge j)}(k)=\begin{cases}1, & \text{if } k\in\mathcal{F};\\ \sum_{i\ge j}\nu^{(i)}(k), & \text{otherwise.}\end{cases}$$

Let I be an interval on the Fermi surface $\mathcal{F}$. Then

$$s = \{k \mid \mathbf{k}'(k) \in I, k \in \operatorname{supp} \nu^{(\geq j-1)}\}$$

is called a sector of length length (I) at scale j. Two different sectors s and s' are called neighbors if $s' \cap s \neq \varnothing$. A sectorization of length $\mathfrak{l}_j$ at scale j is a set of Σ_j of sectors of length $\mathfrak{l}_j$ at scale j that obeys

- the set Σ_j of sectors covers the Fermi surface
- each sector in Σ_j has precisely two neighbors in Σ_j, one to its left and one to its right
- if $s, s' \in \Sigma_j$ are neighbors then $\mathfrak{l}_j/16 \leq \text{length}(s \cap s' \cap \mathcal{F}) \leq \frac{1}{8}\mathfrak{l}_j$.

Observe that there are at most 2 length$(\mathcal{F})/\mathfrak{l}_j$ sectors in Σ_j. In these notes, we fix $\mathfrak{l}_j = 1/M^{j/2}$ and a sectorization Σ_j at scale j.

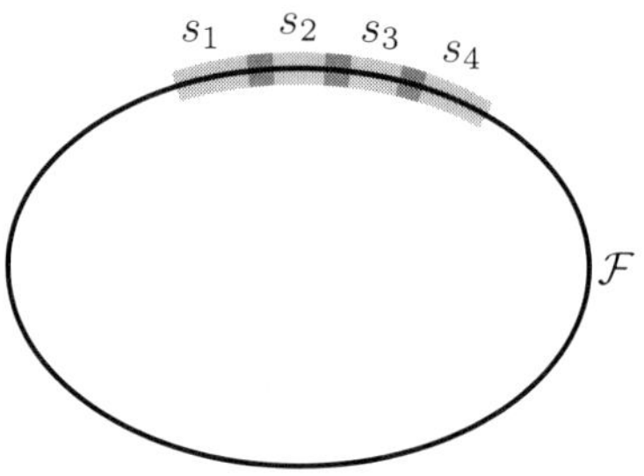

Next we describe how we "sectorize" an interaction

$$W_r = \sum_{\substack{\sigma_i \in \{\uparrow,\downarrow\} \\ \kappa_i \in \{0,1\}}} \int w_r\big((x_1,\sigma_1,\kappa_1),\ldots,(x_r,\sigma_r,\kappa_r)\big)\psi_{\sigma_i}(x_1,\kappa_1)\cdots\psi_{\sigma_r}(x_r,\kappa_r)\prod_{i=1}^{r} dx_i$$

where

$$\psi_{\sigma_i}(x_i) = \psi_{\sigma_i}(x_i,\kappa_i)\mid_{\kappa_i=0}, \qquad \bar{\psi}_{\sigma_i}(x_i) = \psi_{\sigma_i}(x_i,\kappa_i)\mid_{\kappa_i=1}.$$

Let $\mathcal{F}(r,\Sigma_j)$ denote the space of all translation invariant functions

$$f_r\big((x_1,\sigma_1,\kappa_1,s_1),\ldots,(x_r,\sigma_r,\kappa_r,s_r)\big) : (\mathbb{R}^3 \times \{\uparrow,\downarrow\} \times \{0,1\} \times \Sigma_j\})^r \to \mathbb{C}$$

whose Fourier transform, $\hat{f}_r\big((k_1,\sigma_1,\kappa_1,s_1),\ldots,(k_r,\sigma_r,\kappa_r,s_r)\big)$, vanishes unless $k_i \in s_i$. An $f_r \in \mathcal{F}(r,\Sigma_j)$ is said to be a sectorized representative for w_r if

$$\widehat{w}_r\big((k_1,\sigma_1,\kappa_1),\ldots,(k_r,\sigma_r,\kappa_r)\big) = \sum_{\substack{s_i \in \Sigma_j \\ 1\leq i\leq r}} \hat{f}_r\big((k_1,\sigma_1,\kappa_1,s_1),\ldots,(k_r,\sigma_r,\kappa_r,s_r)\big)$$

for all $k_1,\ldots,k_r \in \operatorname{supp}\nu^{(\geq j)}$. It is easy to construct a sectorized representative for w_r by introducing (in momentum space) a partition of unity of $\operatorname{supp}\nu^{(\geq j)}$ subordinate to Σ_j. Furthermore, if f_r is a sectorized representative for w_r, then

$$\begin{aligned}
&\int w_r\big((x_1,\sigma_1,\kappa_1),\ldots,(x_r,\sigma_r,\kappa_r)\big)\psi_{\sigma_1}(x_1,\kappa_1)\cdots\psi_{\sigma_r}(x_r,\kappa_r)\prod_{i=1}^{r} dx_i \\
&= \sum_{\substack{s_i \in \Sigma_j \\ 1\leq i\leq r}} \int f_r\big((x_1,\sigma_1,\kappa_1,s_1),\ldots,(x_r,\sigma_r,\kappa_r,s_r)\big)\psi_{\sigma_1}(x_1,\kappa_1)\cdots\psi_{\sigma_r}(x_r,\kappa_r)\prod_{i=1}^{r} dx_i
\end{aligned}$$

for all $\psi_{\sigma_i}(x_i,\kappa_i)$ "in the support of" $d\mu_{C^{(\ge j)}}$, i.e., provided ψ is integrated out using a Gaussian Grassmann measure whose propagator is supported in $\operatorname{supp}\nu^{(\ge j)}(k)$. Furthermore, by the momentum space support property of f_r,

$$\begin{aligned}\int f_r\big((x_1,\sigma_1,\kappa_1,s_1),\ldots,(x_r,\sigma_r,\kappa_r,s_r)\big)\psi_{\sigma_1}(x_1,\kappa_1)\cdots\psi_{\sigma_r}(x_r,\kappa_r)\prod_{i=1}^r dx_i \\ = \int f_r\big((x_1,\sigma_1,\kappa_1,s_1),\ldots,(x_r,\sigma_r,\kappa_r,s_r)\big)\psi_{\sigma_1}(x_1,\kappa_1,s_1)\cdots\psi_{\sigma_r}(x_r,\kappa_r,s_r)\prod_{i=1}^r dx_i\end{aligned}$$

where

$$\psi_\sigma(x,b,s)=\int d^3y\,\psi_\sigma(y,b,s)\hat{\chi}^{(j)}_s(x-y)$$

and $\hat{\chi}^{(j)}_s$ is the Fourier transform of a function that is identically one on the sector s. This function is chosen shortly before Proposition C.1.

We have expressed the interaction

$$W_r=\sum_{\substack{s_i\in\Sigma_j\\ \sigma_i\in\{\uparrow,\downarrow\}\\ \kappa_i\in\{0,1\}}}\int f_r\big((x_1,\sigma_1,\kappa_1,s_1),\ldots,(x_r,\sigma_r,\kappa_r,s_r)\big)\prod_{i=1}^r\psi_{\sigma_i}(x_i,\kappa_i,s_i)\prod_{i=1}^r dx_i$$

in terms of a sectorized kernel f_r and new "sectorized" fields, $\psi_\sigma(x,\kappa,s)$, that have propagator

$$\begin{aligned}C^{(j)}_{\sigma,\sigma'}\big((x,s),(y,s')\big)&=\int\psi_\sigma(x,0,s)\psi_{\sigma'}(y,1,s')\,d\mu_{C^{(j)}}(\psi)\\ &=\delta_{\sigma,\sigma'}\int\frac{d^3k}{(2\pi)^3}e^{ik\cdot(y-x)}\frac{\nu^{(j)}(k)\chi^{(j)}_s(k)\chi^{(j)}_{s'}(k)}{ik_0-e(\mathbf{k})}.\end{aligned}$$

The momentum space propagator

$$C^{(j)}_{\sigma,\sigma'}(k,s,s')=\delta_{\sigma,\sigma'}\frac{\nu^{(j)}(k)\chi^{(j)}_s(k)\chi^{(j)}_{s'}(k)}{ik_0-e(\mathbf{k})}$$

vanishes unless s and s' are equal or neighbors, is supported in a region of volume $\mathrm{const}\,\mathfrak{l}_j/M^{2j}$ and has supremum bounded by $\mathrm{const}\,M^j$. By an easy variant of Corollary 1.35, the value of F for this propagator is bounded by

$$\mathrm{F}_j\le C_{\mathrm{F}}\Big(\frac{1}{M^{2j}}M^j\mathfrak{l}_j\Big)^{1/2}=C_{\mathrm{F}}\sqrt{\frac{\mathfrak{l}_j}{M^j}}$$

for some constant C_{F}. By Corollary C.3

$$\sup_{x,\sigma,s}\sum_{\sigma,s'}\int d^3y\big|C^{(j)}_{\sigma,\sigma'}\big((x,s),(y,s')\big)\big|\le\mathrm{const}\,M^j$$

so the value of D for this propagator is bounded by

$$\mathrm{D}_j=\frac{1}{\mathfrak{l}_j}M^{2j}.$$

We are now almost ready to define the norm on interactions that replaces the unsectorized norm $\|W\|_j=D_j\|W\|_{\alpha\mathrm{F}_j}$ of the last example. We define a norm on

$\mathcal{F}(r, \Sigma_j)$ by

$$\|f\| = \max_{1\le i\le r} \max_{x_i,\sigma_i,\kappa_i,s_i} \sum_{\substack{\sigma_k,\kappa_k,s_k \\ k\ne i}} \int \prod_{\ell\ne i} dx_\ell \big|f\big((x_1,\sigma_1,\kappa_1,s_1),\ldots,(x_r,\sigma_r,\kappa_r,s_r)\big)\big|$$

and for any translation invariant function

$$w_r\big((x_1,\sigma_1,\kappa_1),\ldots,(x_r,\sigma_r,\kappa_r)\big) \colon (\mathbb{R}^3\times\{\uparrow,\downarrow\}\times\{0,1\})^r \to \mathbb{C}$$

we define

$$\|w_r\|_{\Sigma_j} = \inf\{\|f\| \mid f\in\mathcal{F}(r,\Sigma_j) \text{ a representative for } W\}.$$

The sectorized norm on interactions is

$$\|W\|_{\alpha j} = \mathrm{D}_j \sum_r (\alpha \mathrm{F}_j)^r \|w_r\|_{\Sigma_j} = \sum_r (\alpha C_{\mathrm{F}})^r \mathfrak{l}_j^{(r-2)/2} M^{-j(r-4)/2} \|w_r\|_{\Sigma_j}.$$

PROPOSITION 2.16 (Change of Sectorization). *Let $j' > j \ge 0$. There is a constant C_S, independent of M, j and j', such that for all $r\ge 4$*

$$\|w_r\|_{\Sigma_{j'}} \le \left[C_S \frac{\mathfrak{l}_j}{\mathfrak{l}_{j'}}\right]^{r-3} \|w_r\|_{\Sigma_j}.$$

PROOF. The spin indices σ_i and bar/unbar indices κ_i play no role, so we suppress them. Let $\epsilon>0$ and choose $f_r\in\mathcal{F}(r,\Sigma_j)$ such that

$$w_r(k_1,\ldots,k_r) = \sum_{\substack{s_i\in\Sigma_j \\ 1\le i\le r}} f_r\big((k_1,s_1),\ldots,(k_r,s_r)\big)$$

for all $k_1,\ldots,k_r$ in the support of $\operatorname{supp}\nu^{(\ge j)}$ and

$$\|w_r\|_{\Sigma_j} \ge \|f_r\| - \epsilon.$$

Let

$$1 = \sum_{s'\in\Sigma_{j'}} \chi_{s'}(\mathbf{k}')$$

be a partition of unity of the Fermi curve $\mathcal{F}$ subordinate to the set $\{s'\cap\mathcal{F} \mid s'\in\Sigma_j\}$ of intervals that obeys

$$\sup_{\mathbf{k}'} |\partial_{\mathbf{k}'}^m \chi_{s'}| \le \frac{\mathrm{const}_m}{\mathfrak{l}_{j'}^m}.$$

Fix a function $\varphi\in C_0^\infty\big([0,2)\big)$, independent of j, j' and M, which takes values in $[0,1]$ and which is identically 1 for $0\le x\le 1$. Set

$$\varphi_{j'}(k) = \varphi(M^{2(j'-1)}[k_0^2 + e(\mathbf{k})^2]).$$

Observe that $\varphi_{j'}$ is identically one on the support of $\nu^{(\ge j')}$ and is supported in the support of $\nu^{(\ge j'-1)}$. Define $g_r\in\mathcal{F}(r,\Sigma_{j'})$ by

$$\begin{aligned} g_r\big((k_1,s_1'),\ldots,(k_r,s_r')\big) &= \sum_{\substack{s_\ell\in\Sigma \\ 1\le\ell\le r}} f_r\big((k_1,s_1),\ldots,(k_r,s_r)\big) \prod_{m=1}^r [\chi_{s_m'}(\mathbf{k}_m)\varphi_{j'}(k_m)] \\ &= \sum_{\substack{s_\ell\cap s_\ell'\ne\varnothing \\ 1\le\ell\le r}} f_r\big((k_1,s_1),\ldots,(k_r,s_r)\big) \prod_{m=1}^r [\chi_{s_m'}(\mathbf{k}_m)\varphi_{j'}(k_m)]. \end{aligned}$$

Clearly

$$w_r(k_1,\ldots,k_r) = \sum_{\substack{s'_\ell\in\Sigma_{j'}\\ 1\le\ell\le r}} g_r\big((k_1,s'_1),\ldots,(k_r,s'_r)\big)$$

for all k_ℓ in the support of $\operatorname{supp}\nu^{(\ge j')}$. Define

$$\begin{aligned}\mathrm{Mom}_i(s') = \{(s'_1,\ldots,s'_r)\in\Sigma^r_{j'} \mid s'_i = s' \text{ and there exist } k_\ell\in s'_\ell,\\ 1\le\ell\le r \text{ such that } \textstyle\sum_\ell(-1)^\ell k_\ell = 0\}\end{aligned}$$

Here, I am assuming, without loss of generality, that the even (respectively, odd) numbered legs of w_r are hooked to ψ's (respectively $\bar\psi$'s). Then

$$\|g_r\| = \max_{1\le i\le r}\ \sup_{\substack{x_i\in\mathbb{R}^3\\ s'\in\Sigma_{j'}}}\ \sum_{\mathrm{Mom}_i(s')}\int\prod_{\ell\ne i}dx_\ell\big|g_r\big((x_1,s'_1),\ldots,(x_r,s'_r)\big)\big|.$$

Fix any $1\le i\le r$, $s'\in\Sigma'_j$ and $x_i\in\mathbb{R}^3$. Then

$$\begin{aligned}&\sum_{\mathrm{Mom}_i(s')}\int\prod_{\ell\ne i}dx_\ell\big|g_r\big((x_1,s'_1),\ldots,(x_r,s'_r)\big)\big|\\ &\quad\le\sum_{\mathrm{Mom}_i(s')}\ \sum_{\substack{s_1,\ldots,s_r\\ s_\ell\cap s'_\ell\ne\varnothing}}\int\prod_{\ell\ne i}dx_\ell\,\big|f_r\big((x_1,s_1),\ldots,(x_r,s_r)\big)\big|\max_{s''\in\Sigma_{j'}}\|\widehat{\chi}_{s''}\star\widehat{\varphi}_{j'}\|^r.\end{aligned}$$

By Proposition C.1, with $j=j'$ and $\phi^{(j)}=\widehat{\varphi}_{j'}$, $\max_{s''\in\Sigma_{j'}}\|\widehat{\chi}_{s''}\star\widehat{\varphi}_{j'}\|^r$ is bounded by a constant independent of M, j' and $\mathfrak{l}_{j'}$. Observe that

$$\begin{aligned}&\sum_{\mathrm{Mom}_i(s')}\ \sum_{\substack{s_1,\ldots,s_r\\ s_\ell\cap s'_\ell\ne\varnothing}}\int\prod_{\ell\ne i}dx_\ell\big|f_r\big((x_1,s_1),\ldots,(x_r,s_r)\big)\big|\\ &\qquad\le\sum_{\substack{s_1,\ldots,s_r\\ s_\ell\cap s'_\ell\ne\varnothing}}\ \sum_{\substack{\mathrm{Mom}_i(s')\\ s_\ell\cap s'_\ell\ne\varnothing\\ 1\le\ell\le r}}\int\prod_{\ell\ne i}dx_\ell\big|f_r\big((x_1,s_1),\ldots,(x_r,s_r)\big)\big|.\end{aligned}$$

I will not prove the fact that, for any fixed s_1, …, $s_r\in\Sigma_j$, there are at most $[C'_S\mathfrak{l}_j/\mathfrak{l}_{j'}]^{r-3}$ elements of $\mathrm{Mom}_i(s')$ obeying $s_\ell\cap s'_\ell\ne\varnothing$ for all $1\le\ell\le r$, but I will try to motivate it below. As there are at most two sectors $s\in\Sigma_j$ that intersect s',

$$\begin{aligned}&\sum_{\substack{s_1,\ldots,s_r\\ s_i\cap s'\ne\varnothing}}\ \sum_{\substack{\mathrm{Mom}_i(s')\\ s_\ell\cap s'_\ell\ne\varnothing\\ 1\le\ell\le r}}\int\prod_{\ell\ne i}dx_\ell\,\big|f_r\big((x_1,s_1),\ldots,(x_r,s_r)\big)\big|\\ &\qquad\le 2\left[C'_S\frac{\mathfrak{l}_j}{\mathfrak{l}_{j'}}\right]^{r-3}\sup_{s\in\Sigma_j}\ \sum_{\substack{s_1,\ldots,s_r\\ s_i=s}}\int\prod_{\ell\ne i}dx_\ell\,\big|f_r\big((x_1,s_1),\ldots,(x_r,s_r)\big)\big|\\ &\qquad\le 2\left[C'_S\frac{\mathfrak{l}_j}{\mathfrak{l}_{j'}}\right]^{r-3}\|f_r\|\end{aligned}$$

and

$$\begin{aligned}\|w_r\|_{\Sigma_{j'}}\le\|g_r\|&\le 2\max_{s''\in\Sigma_{j'}}\|\widehat{\chi}_{s''}\star\widehat{\varphi}_{j'}\|^r\left[C'_S\frac{\mathfrak{l}_j}{\mathfrak{l}_{j'}}\right]^{r-3}\|f_r\|\\ &\le\left[C_S\frac{\mathfrak{l}_j}{\mathfrak{l}_{j'}}\right]^{r-3}(\|w_{l,r}\|_{\Sigma_j}+\epsilon)\end{aligned}$$

with $C_S = 2\max_{s''\in\Sigma_{j'}} \|\widehat{\chi}_{s''} \star \widehat{\varphi}_{j'}\|^4 C'_S$.

Now, I will try to motivate the fact that, for any fixed $s_1, \ldots, s_r \in \Sigma_j$, there are at most $[C'_S \mathfrak{l}_j/\mathfrak{l}_{j'}]^{r-3}$ elements of $\mathrm{Mom}_i(s')$ obeying $s_\ell \cap s'_\ell \neq \varnothing$ for all $1 \le \ell \le r$. We may assume that $i = 1$. Then s'_1 must be s'. Denote by I_ℓ the interval on the Fermi curve $\mathcal{F}$ that has length $\mathfrak{l}_j + 2\mathfrak{l}_{j'}$ and is centered on $s_\ell \cap \mathcal{F}$. If $s' \in \Sigma_{j'}$ intersects s_ℓ, then $s' \cap \mathcal{F}$ is contained in I_ℓ. Every sector in $\Sigma_{j'}$ contains an interval of $\mathcal{F}$ of length $\frac{3}{4}\mathfrak{l}_{j'}$ that does not intersect any other sector in $\Sigma_{j'}$. At most $[\frac{4}{3}(\mathfrak{l}_j + 2\mathfrak{l}_{j'})/\mathfrak{l}_{j'}]$ of these "hard core" intervals can be contained in I_ℓ. Thus there are at most $[\frac{4}{3}\mathfrak{l}_j/\mathfrak{l}_{j'} + 3]^{r-3}$ choices for $s'_2, \ldots, s'_{r-2}$.

Fix $s'_1, s'_2, \ldots, s'_{r-2}$. Once s'_{r-1} is chosen, s'_r is essentially uniquely determined by conservation of momentum. But the desired bound on $\mathrm{Mom}_i(s')$ demands more. It says, roughly speaking, that both s'_{r-1} and s'_r are essentially uniquely determined. As k_ℓ runs over s'_ℓ for $1 \le \ell \le r-2$, the sum $\sum_{\ell=1}^{r-2}(-1)^\ell k_\ell$ runs over a small set centered on some point $\mathbf{p}$. In order for $(s'_1, \ldots, s'_r)$ to be in $\mathrm{Mom}_1(s')$, there must exist $\mathbf{k}_{r-1} \in s'_{r-n} \cap \mathcal{F}$ and $\mathbf{k}_r \in s'_r \cap \mathcal{F}$ with $\mathbf{k}_r - \mathbf{k}_{r-1}$ very close to $\mathbf{p}$. But $\mathbf{k}_r - \mathbf{k}_{r-1}$ is a secant joining two points of the Fermi curve $\mathcal{F}$. We have assumed that $\mathcal{F}$ is convex. Consequently, for any given $\mathbf{p} \neq 0$ in $\mathbb{R}^2$ there exist at most two pairs $(\mathbf{k}', \mathbf{q}') \in \mathcal{F}^2$ with $\mathbf{k}' - \mathbf{q}' = \mathbf{p}$. So, if $\mathbf{p}$ is not near the origin, s'_{r-1} and s'_r are almost uniquely determined. If $\mathbf{p}$ is close to zero, then $\sum_{\ell=1}^{r-2}(-1)^\ell k_\ell$ must be close to zero and the number of allowed $s'_r, s'_2, \ldots, s'_{r-2}$ is reduced. □

THEOREM 2.15 MB$_2$. *Suppose $\alpha \ge 2$ and $M \ge \big(\alpha/(\alpha-1)\big)^2 (2C_S(\alpha+1)/\alpha)^{12}$. If $\|W\|_{\alpha,j} \le \frac{1}{3}$ and w_r vanishes for $r \le 4$, then $\|\Omega_{j+1}(W)\|_{\alpha,j+1} \le \|W\|_{\alpha,j}$.*

PROOF. We first verify that $\|W(\Psi+\psi)\|_{\alpha+1,j+1} \le \frac{1}{3}$.

$$\begin{aligned}
&\|W(\Psi+\psi)\|_{\alpha+1,j+1} \\
&\quad = \|W\|_{2(\alpha+1),j+1} \\
&\quad = \sum_r \big(2(\alpha+1)C_{\mathrm{F}}\big)^r \mathfrak{l}_{j+1}^{(r-2)/2} M^{-(j+1)(r-4)/2} \|w_r\|_{\Sigma_{j+1}} \\
&\quad \le \sum_{r\ge 6} \Big(2\frac{\alpha+1}{\alpha}\Big)^r \Big(\frac{\mathfrak{l}_{j+r}}{\mathfrak{l}_j}\Big)^{(r-2)/2} M^{-(r-4)/2} \Big(C_S\frac{\mathfrak{l}_j}{\mathfrak{l}_{j+1}}\Big)^{r-3} (\alpha C_{\mathrm{F}})^r \mathfrak{l}_j^{(r-2)/2} M^{-j(r-4)/2} \|w_r\|_{\Sigma_j} \\
&\quad \le \sum_{r\ge 6} \Big(2C_S\frac{\alpha+1}{\alpha}\Big)^r M^{-(r-4)/2} \Big(\frac{\mathfrak{l}_j}{\mathfrak{l}_{j+1}}\Big)^{r-4/2} (\alpha C_{\mathrm{F}})^r \mathfrak{l}_j^{(r-2)/2} M^{-j(r-4)/2} \|w_r\|_{\Sigma_j} \\
&\quad = \sum_{r\ge 6} \Big(2C_S\frac{\alpha+1}{\alpha} M^{-(1-4/r)/4}\Big)^r (\alpha C_{\mathrm{F}})^r \mathfrak{l}_j^{(r-2)/2} M^{-j(r-4)/2} \|w_r\|_{\Sigma_j} \\
&\quad \le \Big(2C_S\frac{\alpha+1}{\alpha}\Big)^6 \frac{1}{M^{1/2}} \|W\|_{\alpha,j} \le \frac{1}{3}.
\end{aligned}$$

By Corollary 2.7,

$$\begin{aligned}
\|\Omega_{j+1}(W)\|_{\alpha,j+1} &\le \frac{\alpha}{\alpha-1} \|W(\Psi+\psi)\|_{\alpha,j+1} \\
&\le \frac{\alpha}{\alpha-1} \Big(2C_S\frac{\alpha+1}{\alpha}\Big)^6 \frac{1}{M^{1/2}} \|W\|_{\alpha,j} \le \|W\|_{\alpha,j}. \qquad \square
\end{aligned}$$

APPENDIX A

Infinite-Dimensional Grassmann Algebras

To generalize the discussion of Chapter 1 to the infinite-dimensional case we need to add topology. We start with a vector space $\mathcal{V}$ that is an ℓ^1 space. This is not the only possibility. See, for example [**Be**].

Let $\mathcal{I}$ be any countable set. We now generate a Grassmann algebra from the vector space

$$\mathcal{V} = \ell^1(\mathcal{I}) = \left\{ \alpha \colon \mathcal{I} \to \mathbb{C} \;\middle|\; \sum_{i \in \mathcal{I}} |\alpha_i| < \infty \right\}.$$

Equipping $\mathcal{V}$ with the norm $\|\alpha\| = \sum_{i\in\mathcal{I}} |\alpha_i|$ turns it into a Banach space. The algebra will again be an ℓ^1 space. The index set will be $\mathfrak{I}$, the (again countable) set of all finite subsets of $\mathcal{I}$, including the empty set. The Grassmann algebra, with coefficients in $\mathbb{C}$, generated by $\mathcal{V}$ is

$$\mathfrak{A}(\mathcal{I}) = \ell^1(\mathfrak{I}) = \left\{ \alpha \colon \mathfrak{I} \to \mathbb{C} \;\middle|\; \sum_{i \in \mathfrak{I}} |\alpha_i| < \infty \right\}.$$

Clearly $\mathfrak{A} = \mathfrak{A}(\mathcal{I})$ is a Banach space with norm $\|\alpha\| = \sum_{I\in\mathfrak{I}} |\alpha_I|$.

It is also an algebra under the multiplication

$$(\alpha\beta)_{\mathrm{I}} = \sum_{\mathrm{J}\subset\mathrm{I}} \mathrm{sgn}(\mathrm{J}, \mathrm{I}\setminus\mathrm{J})\alpha_{\mathrm{J}}\beta_{\mathrm{I}\setminus\mathrm{J}}.$$

The sign is defined as follows. Fix any ordering of $\mathcal{I}$ and view every $\mathfrak{I} \ni \mathrm{I} \subset \mathcal{I}$ as being listed in that order. Then $\mathrm{sgn}(\mathrm{J}, \mathrm{I}\setminus\mathrm{J})$ is the sign of the permutation that reorders $(\mathrm{J}, \mathrm{I}\setminus\mathrm{J})$ to I. The choice of ordering of $\mathcal{I}$ is arbitrary because the map $\{\alpha_{\mathrm{I}}\} \mapsto \{\mathrm{sgn}(\mathrm{I})\alpha_{\mathrm{I}}\}$, with $\mathrm{sgn}(\mathrm{I})$ being the sign of the permutation that reorders I according to the reordering of $\mathcal{I}$ is an isometric isomorphism. The following bound shows that multiplication is everywhere defined and continuous.

$$\text{(A.1)} \qquad \begin{aligned} \|\alpha\beta\| = \sum_{\mathrm{I}\in\mathfrak{I}} |(\alpha\beta)_{\mathrm{I}}| &= \sum_{\mathrm{I}\in\mathfrak{I}} \left| \sum_{\mathrm{J}\subset\mathrm{I}} \mathrm{sgn}(\mathrm{J}, \mathrm{I}\setminus\mathrm{J})\alpha_{\mathrm{J}}\beta_{\mathrm{I}\setminus\mathrm{J}} \right| \\ &\le \sum_{\mathrm{I}\in\mathfrak{I}} \sum_{\mathrm{J}\subset\mathrm{I}} |\alpha_{\mathrm{J}}|\,|\beta_{\mathrm{I}\setminus\mathrm{J}}| \le \|\alpha\|\,\|\beta\|. \end{aligned}$$

Hence $\mathfrak{A}(\mathcal{I})$ is a Banach algebra with identity $\mathbb{1}_{\mathrm{I}} = \delta_{\mathrm{I},\varnothing}$. In other words, $\mathbb{1}$ is the function on $\mathfrak{I}$ that takes the value one on $\mathrm{I} = \varnothing$ and the value zero on every $\mathrm{I} \neq \varnothing$.

Define, for each $i \in \mathcal{I}$, a_i to be the element of $\mathfrak{A}(\mathcal{I})$ that takes the value 1 on $\mathrm{I} = \{i\}$ and zero otherwise. Also define, for each $\mathrm{I} \in \mathfrak{I}$, a_{I} to be the element of $\mathfrak{A}(\mathcal{I})$ that takes the value 1 on I and zero otherwise. Then

$$a_{\mathrm{I}} = \prod_{i\in\mathrm{I}} a_i$$

where the product is in the order of the ordering of $\mathcal{I}$ and

$$\alpha = \sum_{\mathrm{I} \subset \mathcal{I}} \alpha_{\mathrm{I}} a_{\mathrm{I}}.$$

If $f\colon \mathbb{C} \to \mathbb{C}$ is any function that is defined and analytic in a neighborhood of 0, then the power series $f(\alpha)$ converges for all $\alpha \in \mathfrak{A}(\mathcal{I})$ with $\|\alpha\|$ strictly smaller than the radius of convergence of f since, by (A.1),

$$\|f(\alpha)\| = \left\| \sum_{n=0}^{\infty} \frac{1}{n!} f^{(n)}(0) \alpha^n \right\| \leq \sum_{n=0}^{\infty} \frac{1}{n!} |f^{(n)}(0)| \, \|\alpha\|^n.$$

If f is entire, like the exponential of any polynomial, then $f(\alpha)$ is defined on all of $\mathfrak{A}(\mathcal{I})$.

The following problems give several easy generalizations of the above construction.

PROBLEM A.1. Let $\mathcal{I}$ be any ordered countable set and $\mathfrak{I}$ the set of all finite subsets of $\mathcal{I}$ (including the empty set). Each $\mathrm{I} \in \mathfrak{I}$ inherits an ordering from $\mathcal{I}$. Let $w\colon \mathcal{I} \to (0, \infty)$ be any strictly positive function on $\mathcal{I}$ and set

$$W_{\mathrm{I}} = \prod_{i \in \mathrm{I}} w_i$$

with the convention $W_{\varnothing} = 1$. Define

$$\mathcal{V} = \ell^1(\mathcal{I}, w) = \left\{ \alpha\colon \mathcal{I} \to \mathbb{C} \;\middle|\; \sum_{i \in \mathcal{I}} w_i |\alpha_i| < \infty \right\}$$

and "the Grassmann algebra generated by $\mathcal{V}$"

$$\mathfrak{A}(\mathcal{I}, w) = \ell^1(\mathfrak{I}, W) = \left\{ \alpha : \mathfrak{I} \to \mathbb{C} \;\middle|\; \sum_{\mathrm{I} \in \mathfrak{I}} W_{\mathrm{I}} |\alpha_{\mathrm{I}}| < \infty \right\}.$$

The multiplication is $(\alpha\beta)_{\mathrm{I}} = \sum_{\mathrm{J} \subset \mathrm{I}} \mathrm{sgn}(\mathrm{J}, \mathrm{I} \setminus \mathrm{J}) \alpha_{\mathrm{J}} \beta_{\mathrm{I} \setminus \mathrm{J}}$ where $\mathrm{sgn}(\mathrm{J}, \mathrm{I} \setminus \mathrm{J})$ is the sign of the permutation that reorders $(\mathrm{J}, \mathrm{I} \setminus \mathrm{J})$ to I. The norm $\|\alpha\| = \sum_{\mathrm{I} \in \mathfrak{I}} W_{\mathrm{I}} |\alpha_{\mathrm{I}}|$ turns $\mathfrak{A}(\mathcal{I}, w)$ into a Banach space.

(a) Show that

$$\|\alpha\beta\| \leq \|\alpha\| \, \|\beta\|.$$

(b) Show that if $f\colon \mathbb{C} \to \mathbb{C}$ is any function that is defined and analytic in a neighborhood of 0, then the power series $f(\alpha) = \sum_{n=0}^{\infty} f^{(n)}(0) \alpha^n / n!$ converges for all $\alpha \in \mathfrak{A}$ with $\|\alpha\|$ smaller than the radius of convergence of f.

(c) Prove that $\mathfrak{A}_f(\mathcal{I}) = \{\alpha\colon \mathfrak{I} \to \mathbb{C} \mid \alpha_{\mathrm{I}} = 0 \text{ for all but finitely many I}\}$ is a dense subalgebra of $\mathfrak{A}(\mathcal{I}, w)$.

PROBLEM A.2. Let $\mathcal{I}$ be any ordered countable set and $\mathfrak{I}$ the set of all finite subsets of $\mathcal{I}$. Let

$$\mathfrak{G} = \{\alpha\colon \mathfrak{I} \to \mathbb{C}\}$$

be the set of *all* sequences indexed by $\mathfrak{I}$. Observe that our standard product $(\alpha\beta)_{\mathrm{I}} = \sum_{\mathrm{J} \subset \mathrm{I}} \mathrm{sgn}(\mathrm{J}, \mathrm{I} \setminus \mathrm{J}) \alpha_{\mathrm{J}} \beta_{\mathrm{I} \setminus \mathrm{J}}$ is well-defined on $\mathfrak{G}$—for each $\mathrm{I} \in \mathfrak{I}$, $\sum_{\mathrm{J} \subset \mathrm{I}}$ is a finite sum. We now define, for each integer n, a norm on (a subset of) $\mathfrak{G}$ by

$$\|\alpha\|_n = \sum_{\mathrm{I} \in \mathfrak{I}} 2^{n|\mathrm{I}|} |\alpha_{\mathrm{I}}|.$$

It is defined for all $\alpha \in \mathfrak{G}$ for which the series converges. Observe that this is precisely the norm of Problem A.1 with $w_i = 2^n$ for all $i \in \mathcal{I}$. Also observe that, if $m < n$, then $\|\alpha\|_m \le \|\alpha\|_n$. Define

$$\mathfrak{A}_\cap = \{\alpha \in \mathfrak{G} \mid \|\alpha\|_n < \infty \text{ for all } n \in \mathbb{Z}\}$$
$$\mathfrak{A}_\cup = \{\alpha \in \mathfrak{G} \mid \|\alpha\|_n < \infty \text{ for some } n \in \mathbb{Z}\}.$$

(a) Prove that if $\alpha,\ \beta \in \mathfrak{A}_\cap$ then $\alpha\beta \in \mathfrak{A}_\cap$.
(b) Prove that if $\alpha,\ \beta \in \mathfrak{A}_\cup$ then $\alpha\beta \in \mathfrak{A}_\cup$.
(c) Prove that if $f(z)$ is an entire function and $\alpha \in \mathfrak{A}_\cap$, then $f(\alpha) \in \mathfrak{A}_\cap$.
(d) Prove that if $f(z)$ is analytic at the origin and $\alpha \in \mathfrak{A}_\cup$ has $|\alpha_\emptyset|$ strictly smaller than the radius of convergence of f, then $f(\alpha) \in \mathfrak{A}_\cup$.

Before moving on to integration, we look at some examples of Grassmann algebras generated by ordinary functions or distributions on $\mathbb{R}^d$. In these algebras we can have elements like

$$\mathcal{A}(\psi,\bar\psi) = -\sum_{\sigma\in\mathfrak{S}} \int \frac{d^{d+1}k}{(2\pi)^{d+1}} \left(ik_0 - \left(\frac{\mathbf{k}^2}{2m} - \mu\right)\right) \bar\psi_{k,\sigma}\psi_{k,\sigma}$$
$$-\frac{\lambda}{2} \sum_{\sigma,\sigma'\in\mathfrak{S}} \int \prod_{i=1}^{4} \frac{d^{d+1}}{(2\pi)^{d+1}} \delta(k_1+k_2-k_3-k_4) \bar\psi_{k_1,\sigma}\psi_{k_3,\sigma}\hat u(\mathbf{k}_1-\mathbf{k}_3)\bar\psi_{k_2,\sigma'}\psi_{k_4,\sigma'}$$

from Section 1.5. that look like they are in an algebra with uncountable dimension. But they really aren't.

Example A.1 (Functions and distributions on $\mathbb{R}^d/L\mathbb{Z}^d$). There is a natural way to express the space of smooth functions on the torus $\mathbb{R}^d/L\mathbb{Z}^d$ (i.e., smooth functions on $\mathbb{R}^d$ that are periodic with respect to $L\mathbb{Z}^d$) as a space of sequences: Fourier series. The same is true for the space of distributions on $\mathbb{R}^d/L\mathbb{Z}^d$. By definition, a distribution on $\mathbb{R}^d/L\mathbb{Z}^d$ is a map

$$f\colon C^\infty(\mathbb{R}^d/L\mathbb{Z}^d) \longrightarrow \mathbb{C}$$
$$h \longmapsto \langle f, h\rangle$$

that is linear in h and is continuous in the sense that there exist constants $C_f \in \mathbb{R}$, $\nu_f \in \mathbb{N}$ such that

$$|\langle f,h\rangle| \le \sup_{\mathbf{x}\in\mathbb{R}^d/L\mathbb{Z}^d} C_f \left| \prod_{j=1}^{d} \left(1 - \frac{d^2}{d\mathbf{x}_j^2}\right)^{\nu_f} h(\mathbf{x})\right| \tag{A.2}$$

for all $h \in C^\infty$. Each L^1 function $f(\mathbf{x})$ on $\mathbb{R}^d/L\mathbb{Z}^d$ is identified with the distribution

$$\langle f,h\rangle = \int_{\mathbb{R}^d/L\mathbb{Z}^d} d\mathbf{x}\, f(\mathbf{x})h(\mathbf{x}) \tag{A.3}$$

which has $C_f = \|f\|_{L^1}$, $\nu_f = 0$. The delta "function" supported at $\mathbf{x}$ is really the distribution

$$\langle \delta_{\mathbf{x}}, h\rangle = h(\mathbf{x})$$

and has $C_{\delta_{\mathbf{x}}} = 1$, $\nu_{\delta_{\mathbf{x}}} = 0$.

Let $\gamma \in \mathbb{Z}$. We now define a Grassmann algebra $\mathfrak{A}_\gamma(\mathbb{R}^d/L\mathbb{Z}^d)$, $\gamma \in \mathbb{Z}$. It is the Grassmann algebra of Problem A.1 with index set

$$\mathcal{I} = \frac{2\pi}{L}\mathbb{Z}^d$$

and weight function

$$w_{\mathbf{p}}^{\gamma} = \prod_{j=1}^{d} (1 + \mathbf{p}_j^2)^{\gamma}.$$

To each distribution f, we associate the sequence

$$\tilde{f}_{\mathbf{p}} = \langle f, e^{-i\langle \mathbf{p}, \mathbf{x} \rangle} \rangle$$

indexed by $\mathcal{I}$. In the event that f is given by integration against and L^1 function, then

$$\tilde{f}_{\mathbf{p}} = \int_{\mathbb{R}^d / L\mathbb{Z}^d} d\mathbf{x}\, f(\mathbf{x}) e^{-i\langle \mathbf{p}, \mathbf{x} \rangle}$$

is the usual Fourier coefficient. For example, let $\mathbf{q} \in (2\pi/L)\mathbb{Z}^d$ and $f(\mathbf{x}) = e^{i\langle \mathbf{q}, \mathbf{x} \rangle}/L^d$. Then

$$\tilde{f}_{\mathbf{p}} = \begin{cases} 1, & \text{if } \mathbf{p} = \mathbf{q} \\ 0, & \text{if } \mathbf{p} \neq \mathbf{q} \end{cases}$$

and

$$\sum_{\mathbf{p} \in (2\pi/L)\mathbb{Z}^d} w_{\mathbf{p}}^{\gamma} |\tilde{f}_{\mathbf{p}}| = w_{\mathbf{q}}^{\gamma}$$

is finite for all γ. Call this sequence (the sequence whose $\mathbf{p}$th entry is one when $\mathbf{p} = \mathbf{q}$ and zero otherwise) $\psi_{\mathbf{q}}$. We have just shown that $\psi_{\mathbf{q}} \in \mathcal{V}_{\gamma}(\mathbb{R}^d/L\mathbb{Z}^d) \subset \mathfrak{A}_{\gamma}(\mathbb{R}^d/L\mathbb{Z}^d)$ for all $\gamma \in \mathbb{Z}$. Define, for each $\mathbf{x} \in \mathbb{R}^d/L\mathbb{Z}^d$,

$$\psi(\mathbf{x}) = \frac{1}{L^d} \sum_{\mathbf{q} \in (2\pi/L)\mathbb{Z}^d} e^{i\langle \mathbf{q}, \mathbf{x} \rangle} \psi_{\mathbf{q}}.$$

The $\mathbf{p}$th entry in the sequence for $\psi(\mathbf{x})$ is $(1/L^d) \sum_{\mathbf{q}} e^{i\langle \mathbf{q}, \mathbf{x} \rangle} \delta_{\mathbf{p},\mathbf{q}} = (1/L^d) e^{i\langle \mathbf{p}, \mathbf{x} \rangle}$. As

$$\frac{1}{L^d} \sum_{\mathbf{p} \in (2\pi/L)\mathbb{Z}^d} w_{\mathbf{p}}^{\gamma} |e^{i\langle \mathbf{p}, \mathbf{x} \rangle}| = \frac{1}{L^d} \sum_{\mathbf{p} \in (2\pi/L)\mathbb{Z}^d} w_{\mathbf{p}}^{\gamma}$$

converges for all $\gamma < -\frac{1}{2}$, $\psi(\mathbf{x}) \in \mathcal{V}_{\gamma}(\mathbb{R}^d/L\mathbb{Z}^d) \subset \mathfrak{A}_{\gamma}(\mathbb{R}^d/L\mathbb{Z}^d)$ for all $\gamma < -\frac{1}{2}$.

By (A.2), for each distribution f, there are constants C_f and ν_f such that

$$|\tilde{f}_{\mathbf{p}}| = |\langle f, e^{-i\langle \mathbf{p}, \mathbf{x} \rangle} \rangle| \leq \sup_{\mathbf{x} \in \mathbb{R}^d/L\mathbb{Z}^d} C_f \left| \prod_{j=1}^{d} \left(1 - \frac{d^2}{d\mathbf{x}_j^2}\right)^{\nu_f} e^{-i\langle \mathbf{p}, \mathbf{x} \rangle} \right| = C_f \prod_{j=1}^{d} (1 + \mathbf{p}_j^2)^{\nu_f}.$$

Because

$$\sum_{\mathbf{p} \in (2\pi/L)\mathbb{Z}^d} \prod_{j=1}^{d} (1 + \mathbf{p}_j^2)^{\gamma}$$

converges for all $\gamma < -\frac{1}{2}$, the sequence $\{\tilde{f}_{\mathbf{p}} \mid \mathbf{p} \in (2\pi/L)\mathbb{Z}^d\}$ is in $\mathcal{V}_{\gamma}(\mathbb{R}^d/L\mathbb{Z}^d) = \ell^1(\mathcal{I}, w^{\mathcal{V}}) \subset \mathfrak{A}_{\gamma}(\mathbb{R}^d/L\mathbb{Z}^d)$, for all $\gamma < -\nu_f - \frac{1}{2}$. Thus, every distribution is in $\mathcal{V}_{\gamma}$ for some γ (often negative). On the other hand, by Problem A.4 below, a distribution is given by integration against a C^{∞} function (i.e., is of the form (A.3) for some periodic C^{∞} function $f(\mathbf{x})$) if and only if $\tilde{f}_{\mathbf{p}}$ decays faster than the inverse of any polynomial in $\mathbf{p}$ and then it (or rather $\tilde{f}$) is in $\mathcal{V}_{\gamma}$ for every γ.

PROBLEM A.3. (a) Let $f(x) \in C^\infty(\mathbb{R}^d/L\mathbb{Z}^d)$. Define

$$\tilde{f}_{\mathbf{p}} = \int_{\mathbb{R}^d/L\mathbb{Z}^d} d\mathbf{x}\, f(\mathbf{x}) e^{-i\langle \mathbf{p},\mathbf{x}\rangle}.$$

Prove that for every $\gamma \in \mathbb{N}$, there is a constant $C_{f,\gamma}$ such that

$$|\tilde{f}_{\mathbf{p}}| \le C_{f,\gamma} \prod_{j=1}^{d} (1+\mathbf{p}_j^2)^{-\gamma} \quad \text{for all } \mathbf{p} \in \frac{2\pi}{L}\mathbb{Z}^d.$$

(b) Let f be a distribution on $\mathbb{R}^d/L\mathbb{Z}^d$. For any $h \in C^\infty(\mathbb{R}^d/L\mathbb{Z}^d)$, set

$$h_R(\mathbf{x}) = \frac{1}{L^d} \sum_{\substack{\mathbf{p}\in(2\pi/L)\mathbb{Z}^d \\ |\mathbf{p}|<R}} e^{i\langle \mathbf{p},\mathbf{x}\rangle} \tilde{h}_{\mathbf{p}}, \quad \text{where } \tilde{h}_{\mathbf{p}} = \int_{\mathbb{R}^d/L\mathbb{Z}^d} d\mathbf{x}\, h(\mathbf{x}) e^{-i\langle \mathbf{p},\mathbf{x}\rangle}.$$

Prove that

$$\langle f, h\rangle = \lim_{R\to\infty} \langle f, h_R\rangle.$$

PROBLEM A.4. Let f be a distribution on $\mathbb{R}^d/L\mathbb{Z}^d$. Suppose that, for each $\gamma \in \mathbb{Z}$, there is a constant $C_{f,\gamma}$ such that

$$|\langle f, e^{-i\langle \mathbf{p},\mathbf{x}\rangle}\rangle| \le C_{f,\gamma} \prod_{j=1}^{d} (1+\mathbf{p}_j^2)^{-\gamma} \quad \text{for all } \mathbf{p} \in \frac{2\pi}{L}\mathbb{Z}^d.$$

Prove that there is a C^∞ function $F(\mathbf{x})$ on $\mathbb{R}^d/L\mathbb{Z}^d$ such that

$$\langle f, h\rangle = \int_{\mathbb{R}^d/L\mathbb{Z}^d} d\mathbf{x}\, F(\mathbf{x}) h(\mathbf{x}).$$

EXAMPLE A.2 (Functions and distributions on $\mathbb{R}^d$). Example A.1 exploited the basis $\{e^{i\langle \mathbf{p},\mathbf{x}\rangle} \mid \mathbf{p} \in (2\pi/L)\mathbb{Z}^d\}$ of $L^2(\mathbb{R}^d/L\mathbb{Z}^d)$. This is precisely the basis of eigenfunctions of $\prod_{j=1}^d (1 - d^2/dx_j^2)$. If instead we use the basis of $L^2(\mathbb{R}^d)$ given by the eigenfunctions of $\prod_{j=1}^d (x_j^2 - d^2/dx_j^2)$ we get a very useful identification between tempered distributions and functions on the countable index set

$$\mathcal{I} = \{i = (i_1, \ldots, i_d) \mid i_j \in \mathbb{Z}^{\ge 0}\}$$

that we now briefly describe. For more details see [**RS**, Appendix to V.2].

Define the differential operators

$$\mathbb{A} = \frac{1}{\sqrt{2}}\left(x + \frac{d}{dx}\right) \qquad \mathbb{A}^\dagger = \frac{1}{\sqrt{2}}\left(x - \frac{d}{dx}\right)$$

and the Hermite functions

$$h_\ell(x) = \begin{cases} \frac{1}{\pi^{1/4}} e^{-x^2/2}, & \ell = 0 \\ \frac{1}{\sqrt{\ell!}} (\mathbb{A}^\dagger)^\ell h_0(x), & \ell > 0 \end{cases}$$

for $\ell \in \mathbb{Z}^{\ge 0}$.

PROBLEM A.5. Prove that

(a) $\mathbb{A}\mathbb{A}^\dagger f = \mathbb{A}^\dagger \mathbb{A} f + f$
(b) $\mathbb{A} h_0 = 0$
(c) $\mathbb{A}^\dagger \mathbb{A} h_\ell = \ell h_\ell$
(d) $\langle h_\ell, h_{\ell'}\rangle = \delta_{\ell,\ell'}$. Here $\langle f, g\rangle = \int \bar{f}(x) g(x)\, dx$.

(e) $x^2 - d^2/dx^2 = 2\mathbb{A}^\dagger\mathbb{A} + 1$.

Define the multidimensional Hermite functions

$$h_i(x) = \prod_{j=1}^{d} h_{i_j}(x_j)$$

for $i \in \mathcal{I}$. They form an orthonormal basis for $L^2(\mathbb{R}^d)$ and obey

$$\prod_{j=1}^{d}\left(x_j^2 - \frac{d^2}{dx_j^2}\right)^\gamma h_i = \prod_{j=1}^{d}(1+2i_j)^\gamma h_i.$$

By definition, Schwartz space $\mathcal{S}(\mathbb{R}^d)$ is the set of C^∞ functions on $\mathbb{R}^d$ all of whose derivatives decay faster than any polynomial. That is, a function $f(x)$ is in $\mathcal{S}(\mathbb{R}^d)$, if it is C^∞ and

$$\sup_x \left|\prod_{j=1}^{d}\left(x_j^2 - \frac{d^2}{dx_j^2}\right)^\gamma f(x)\right| < \infty$$

for all $\gamma \in \mathbb{Z}^{\geq 0}$. A tempered distribution on $\mathbb{R}^d$ is a map

$$\begin{aligned} f\colon \mathcal{S}(\mathbb{R}^d) &\longrightarrow \mathbb{C} \\ h &\longmapsto \langle f, h\rangle \end{aligned}$$

that is linear in h and is continuous in the sense that there exist constants $C \in \mathbb{R}$, $\gamma \in \mathbb{N}$ such that

$$|\langle f, h\rangle| \leq \sup_{x\in\mathbb{R}^d} C\left|\prod_{j=1}^{d}\left(x_j^2 - \frac{d^2}{dx_j^2}\right)^\gamma h(x)\right|.$$

To each tempered distribution we associate the function

$$\breve{f}_i = \langle f, h_i\rangle$$

on $\mathcal{I}$. In the event that the distribution f is given by integration against an L^2 function, that is,

$$\langle f, h\rangle = \int F(x)h(x)\, d^d x \tag{A.4}$$

for some L^2 function, $F(x)$, then

$$\breve{f}_i = \int F(x)h_i(x)\, d^d x.$$

The distribution f is Schwartz class (meaning that one may choose the $F(x)$ of (A.4) in $\mathcal{S}(\mathbb{R}^d)$) if and only if $\breve{f}_i$ decays faster than any polynomial in i. For any distribution, $\breve{f}_i$ is bounded by some polynomial in i. So, if we define

$$w_i^\gamma = \prod_{j=1}^{d}(1+2i_j)^\gamma$$

then every Schwartz class function (or rather its $\breve{f}$) is in $\mathcal{V}_\gamma = \ell^1(\mathcal{I}, w^\gamma)$ for every γ and every tempered distribution is in $\mathcal{V}_\gamma$ for some γ. The Grassmann algebra generated by $\mathcal{V}_\gamma$ is called $\mathfrak{A}_\gamma(\mathbb{R}^d)$. In particular, since the Hermite functions are continuous and uniformly bounded [**AS**, p. 787], every delta function $\delta(x - x_0)$ and indeed every finite measure has $|\breve{f}_i| \leq$ const. Thus, since $\sum_{i\in\mathbb{Z}_{\geq 0}}(1+i)^\gamma$

converges for all $\gamma < -1$, all finite measures are in $\mathcal{V}_\gamma$ for all $\gamma < -1$. The sequence representation of Schwartz class functions and of tempered distributions is discussed in more detail in [**RS**, Appendix to V.3].

PROBLEM A.6. Show that the constant function $f = 1$ is in $\mathcal{V}_\gamma$ for all $\gamma < -1$. Hint: the Fourier transform of h_ℓ is $(-i)^\ell h_\ell$.

Infinite-Dimensional Grassmann Integrals. Once again, let $\mathcal{I}$ be any countable set, $\mathfrak{I}$ the set of all finite subsets of $\mathcal{I}$, $\mathcal{V} = \ell^1(\mathcal{I})$ and $\mathfrak{A} = \ell^1(\mathfrak{I})$. We now define some linear functionals on $\mathfrak{A}$, Of course the infinite-dimensional analogue of $\int \cdot\, da_D \cdots da_1$ will not make sense. But we can define the analogue of the Grassmann Gaussian integral, $\int \cdot\, d\mu_S$, at least for suitable S.

Let $S\colon \mathcal{I} \times \mathcal{I} \to \mathbb{C}$ be an infinite matrix. Recall that, for each $i \in \mathcal{I}$, a_i is the element of $\mathfrak{A}$ that takes the value 1 on $\mathrm{I} = \{i\}$ and zero otherwise, and, for each $\mathrm{I} \in \mathfrak{I}$, a_{I} is the element of $\mathfrak{A}$ that takes the value 1 on I and zero otherwise. By analogy with the finite-dimensional case, we wish to define a liner functional on $\mathfrak{A}$ by

$$\int \prod_{j=1}^{p} a_{i_j}\, d\mu_S = \mathrm{Pf}[S_{i_j,i_k}]_{1\le j,k\le p}.$$

We now find conditions under which this defines a bounded linear functional. Any $A \in \mathfrak{A}$ can be written in the form

$$A = \sum_{p=0}^{\infty} \sum_{\substack{\mathrm{I}\subset\mathcal{I}\\ |\mathrm{I}|=p}} \alpha_{\mathrm{I}} a_{\mathrm{I}}.$$

We want the integral of this A to be

$$\begin{aligned}\int A\, d\mu_S &= \sum_{p=0}^{\infty} \sum_{\substack{\mathrm{I}\subset\mathcal{I}\\ |\mathrm{I}|=p}} \alpha_{\mathrm{I}} \int a_{\mathrm{I}}\, d\mu_S \\ &= \sum_{p=0}^{\infty} \sum_{\substack{\mathrm{I}\subset\mathcal{I}\\ |\mathrm{I}|=p}} \alpha_{\mathrm{I}}\, \mathrm{Pf}[S_{\mathrm{I}_j,\mathrm{I}_k}]_{1\le j,k\le p}\end{aligned} \tag{A.5}$$

where $\mathrm{I} = \{\mathrm{I}_1, \ldots, \mathrm{I}_p\}$ with $\mathrm{I}_1 < \cdots < \mathrm{I}_p$ in the ordering of $\mathcal{I}$.

We now hypothesize that there is a number of C_S such that

$$\sup_{\substack{\mathrm{I}\subset\mathcal{I}\\ |\mathrm{I}|=p}} |\,\mathrm{Pf}[S_{\mathrm{I}_j,\mathrm{I}_k}]_{1\le j,k\le p}| \le C_S^p. \tag{A.6}$$

Such a bound is the conclusion of the following simple lemma. The bound proven in this lemma is not tight enough to be of much use in quantum field theory applications. A much better bound for those applications is proven in Corollary 1.35.

LEMMA A.3. *Let S be a bounded linear operator on $\ell^2(\mathcal{I})$. Then*

$$\sup_{\substack{\mathrm{I}\subset\mathcal{I}\\ |\mathrm{I}|=p}} |\,\mathrm{Pf}[S_{\mathrm{I}_j,\mathrm{I}_k}]_{1\le j,k\le p}| \le \|S\|^{p/2}.$$

PROOF. Recall that Hadamard's inequality (Problem 1.24) bounds a determinant by the product of the lengths of its columns. The kth column of the matrix

$[S_{\mathrm{I}_j,\mathrm{I}_k}]_{1\le j,k\le p}$ has length

$$\sqrt{\sum_{j=1}^{p}|S_{\mathrm{I}_j,\mathrm{I}_k}|^2} \le \sqrt{\sum_{j\in\mathcal{I}}|S_{j,\mathrm{I}_k}|^2}.$$

But the vector $(S_{j,\mathrm{I}_k})_{j\in\mathcal{I}}$ is precisely the image under S of the $\mathrm{I}_k^{\mathrm{th}}$ standard unit basis vector, $(\delta_{j,\mathrm{I}_k})_{j\in\mathcal{I}}$, in $\ell^2(\mathcal{I})$ and hence has length at most $\|S\|$. Hence

$$\det[S_{\mathrm{I}_j,\mathrm{I}_k}]_{1\le j,k\le p} \le \prod_{k=1}^{p}\sqrt{\sum_{j\in\mathcal{I}}|S_{j,\mathrm{I}_k}|^2} \le \|S\|^p.$$

By Proposition 1.18.d,

$$|\operatorname{Pf}[S_{\mathrm{I}_j,\mathrm{I}_k}]_{1\le j,k\le p}| \le |\det[S_{\mathrm{I}_j,\mathrm{I}_k}]_{1\le j,k\le p}|^{1/2} \le \|S\|^{p/2}. \qquad \square$$

Under the hypothesis (A.6), the integral (A.5) is bounded by

$$\left|\int A\,d\mu_C\right| \le \sum_{p=0}^{\infty}\sum_{\substack{\mathrm{I}\in\mathfrak{I}\\|\mathrm{I}|=p}}|\alpha_{\mathrm{I}}|C_S^p \le \|A\| \tag{A.7}$$

provided $C_S \le 1$. Thus, if $C_S \le 1$, the map $A \in \mathfrak{A} \mapsto \int A\,d\mu_S$ is a bounded linear map with norm at most one.

The requirement that C_S be less than *one* is not crucial. Define, for each $n\in\mathbb{N}$ the Grassmann algebra

$$\mathfrak{A}_n = \left\{\alpha\colon \mathfrak{I}\to\mathbb{C} \;\middle|\; \|\alpha\|_n = \sum_{\mathrm{I}\in\mathrm{I}} 2^{n|\mathrm{I}|}|\alpha_{\mathrm{I}}| < \infty\right\}.$$

If $2^n \ge C_S$, then (A.7) shows that $A\in\mathfrak{A}\mapsto\int A\,d\mu_S$ is a bounded linear functional on $\mathfrak{A}_n$. Alternatively, we can view $\int\cdot\,d\mu_S$ as an unbounded linear functional on $\mathfrak{A}$ with domain of definition the dense subalgebra $\mathfrak{A}_n$ of $\mathfrak{A}$.

APPENDIX B

Pfaffians

DEFINITION B.1. Let $S = (S_{ij})$ be a complex $n \times n$ matrix with $n = 2m$ even. By definition, the Pfaffian $\mathrm{Pf}(S)$ of S is

$$\mathrm{Pf}(S) = \frac{1}{2^m m!} \sum_{i_1,\ldots,i_n=1}^{n} \varepsilon^{i_1\ldots i_n} S_{i_1 i_2} \cdots S_{i_{n-1} i_n} \tag{B.1}$$

where

$$\varepsilon^{i_1\ldots i_n} = \begin{cases} 1, & \text{if } i_1,\ldots,i_n \text{ is an even permutation of } 1,\ldots,n \\ -1, & \text{if } i_1,\ldots,i_n \text{ is an odd permutation of } 1,\ldots,n \\ 0, & \text{if } i_1,\ldots,i_n \text{ are not all distinct.} \end{cases}.$$

By convention, the Pfaffian of a matrix of odd order is zero.

PROBLEM B.1. Let $T = (T_{ij})$ be a complex $n \times n$ matrix with $n = 2m$ even and let $S = \frac{1}{2}(T - T^t)$ be its skew symmetric part. Prove that $\mathrm{Pf}\, T = \mathrm{Pf}\, S$.

PROBLEM B.2. Let $S = \left(\begin{smallmatrix} 0 & S_{12} \\ S_{21} & 0 \end{smallmatrix}\right)$ with $S_{21} = -S_{12} \in \mathbb{C}$. Show that $\mathrm{Pf}\, S = S_{12}$.

For the rest of this appendix, we assume that S is a skew symmetric matrix. Let

$$\mathcal{P}_m = \big\{\{(k_1,\ell_1),\ldots,(k_m,\ell_m)\} \bigm| \{k_1,\ell_1,\ldots,k_m,\ell_m\} = \{1,\ldots,2m\}\big\}$$

be the set of all partitions of $\{1,\ldots,2m\}$ into m disjoint ordered pairs. Observe that, the expression $\varepsilon^{k_1\ell_1\ldots k_m\ell_m} S_{k_1,\ell_1} \cdots S_{k_m,\ell_m}$ is invariant under permutations of the pairs in the partition $\{(k_1,\ell_1),\ldots,(k_m,\ell_m)\}$, since

$$\varepsilon^{k_{\pi(1)}\ell_{\pi(1)}\cdots k_{\pi(m)}\ell_{\pi(m)}} = \varepsilon^{k_1\ell_1\ldots k_m\ell_m}$$

for any permutation π. Thus,

$$\mathrm{Pf}(S) = \frac{1}{2^m} \sum_{\mathbf{P}\in\mathcal{P}_m} \varepsilon^{k_1\ell_1\ldots k_m\ell_m} = S_{k_1\ell_1} \cdots S_{k_m\ell_m}. \tag{B.1$'$}$$

Let

$$\mathcal{P}_m^< = \big\{\{(k_1,\ell_1),\ldots,(k_m,\ell_m)\} \in \mathcal{P}_m \bigm| k_i < \ell_i \text{ for all } 1 \le i \le m\big\}.$$

Because S is skew symmetric

$$\varepsilon^{k_1\ell_1\ldots k_i\ell_i\ldots k_m\ell_m} S_{k_i\ell_i} = \varepsilon^{k_1\ell_1\ldots \ell_i k_i\ldots k_m\ell_m} S_{\ell_i k_i}$$

so that

$$\mathrm{Pf}(S) = \sum_{\mathbf{P}\in\mathcal{P}_m^<} \varepsilon^{k_1\ell_1\ldots k_m\ell_m} S_{k_1\ell_1} \cdots S_{k_m\ell_m}. \tag{B.1$''$}$$

PROBLEM B.3. Let $\alpha_1, \ldots, \alpha_r$ be complex numbers and let S be the $2r \times 2r$ skew symmetric matrix

$$S = \bigoplus_{m=1}^{r} \begin{bmatrix} 0 & \alpha_m \\ -\alpha_m & 0 \end{bmatrix}.$$

All matrix elements of S are zero except for r 2×2 blocks running down the diagonal. For example, if $r = 2$,

$$S = \begin{bmatrix} 0 & \alpha_1 & 0 & 0 \\ -\alpha_1 & 0 & 0 & 0 \\ 0 & 0 & 0 & \alpha_2 \\ 0 & 0 & -\alpha_2 & 0 \end{bmatrix}$$

Prove that $\mathrm{Pf}(S) = \alpha_1\alpha_2 \cdots \alpha_r$.

PROPOSITION B.2. *Let $S = (S_{ij})$ be a skew symmetric matrix of even order $n = 2m$.*

(a) *Let π be a permutation and set $S^\pi = (S_{\pi(i)\pi(j)})$. Then,*

$$\mathrm{Pf}(S^\pi) = \mathrm{sgn}(\pi)\,\mathrm{Pf}(S).$$

(b) *For all $1 \le k \ne \ell \le n$, let $M_{k\ell}$ be the matrix obtained from S by deleting rows k and ℓ and columns k and ℓ. Then*

$$\mathrm{Pf}(S) = \sum_{\ell=1}^{n} \mathrm{sgn}(k-\ell)(-1)^{k+\ell} S_{k\ell}\,\mathrm{Pf}(M_{k\ell}).$$

In particular,

$$\mathrm{Pf}(S) = \sum_{\ell=2}^{n} (-1)^{\ell} S_{1\ell}\,\mathrm{Pf}(M_{1\ell}).$$

PROOF. (a) Recall that π is a fixed permutation of $1, \ldots, n$. Making the change of summation variables $j_1 = \pi(i_1), \ldots, j_n = \pi(i_n)$,

$$\begin{aligned}
\mathrm{Pf}(S^\pi) &= \frac{1}{2^m m!} \sum_{i_1,\ldots,i_n=1}^{n} \varepsilon^{i_1\ldots i_n} S_{\pi(i_1)\pi(i_2)} \cdots S_{\pi(i_{n-1})\pi(i_n)} \\
&= \frac{1}{2^m m!} \sum_{j_1,\ldots,j_n=1}^{n} \varepsilon^{\pi^{-1}(j_1)\ldots\pi^{-1}(j_n)} S_{j_1j_2} \cdots S_{j_{n-1}j_n} \\
&= \mathrm{sgn}(\pi^{-1}) \frac{1}{2^m m!} \sum_{j_1,\ldots,j_n=1}^{n} \varepsilon^{j_1\ldots j_n} S_{j_1j_2} \cdots S_{j_{n-1}j_n} \\
&= \mathrm{sgn}(\pi^{-1})\,\mathrm{Pf}(S) = \mathrm{sgn}(\pi)\,\mathrm{Pf}(S).
\end{aligned}$$

(b) Fix $1 \le k \le n$. Let $\mathbf{P} = \{(k_1, \ell_1), \ldots, (k_m, \ell_m)\}$ be a partition in $\mathcal{P}_m$. We may assume, by reindexing the pairs, that $k = k_1$ or $k = \ell_1$. Then,

$$\mathrm{Pf}(S) = \frac{1}{2^m} \sum_{\substack{\mathbf{P}\in\mathcal{P}_m \\ k_1=k}} \varepsilon^{k\ell_1\ldots k_m\ell_m} S_{k\ell_1} \cdots S_{k_m\ell_m} + \frac{1}{2^m} \sum_{\substack{\mathbf{P}\in\mathcal{P}_m \\ \ell_1=k}} \varepsilon^{k_1k\ldots k_m\ell_m} S_{k_1k} \cdots S_{k_m\ell_m}.$$

By antisymmetry and a change of summation variable,

$$\sum_{\substack{\mathbf{P}\in\mathcal{P}_m \\ \ell_1=k}} \varepsilon^{k_1k\ldots k_m\ell_m} S_{k_1k} \cdots S_{k_m\ell_m} = \sum_{\substack{\mathbf{P}\in\mathcal{P}_m \\ k_1=k}} \varepsilon^{k\ell_1\ldots k_m\ell_m} S_{k\ell_1} \cdots S_{k_m\ell_m}.$$

Thus,

$$\begin{aligned}\mathrm{Pf}(S) &= \frac{2}{2^m} \sum_{\substack{\mathbf{P}\in\mathcal{P}_m \\ k_1=k}} \varepsilon^{k\ell_1\ldots k_m\ell_m} S_{k\ell_1}\ldots S_{k_m\ell_m} \\ &= \frac{2}{2^m} \sum_{\ell=1}^{n} \sum_{\substack{\mathbf{P}\in\mathcal{P}_m \\ k_1=k,\ell_1=\ell}} \varepsilon^{k\ell\ldots k_m\ell_m} S_{k\ell}\cdots S_{k_m\ell_m}.\end{aligned}$$

Extracting $S_{k\ell}$ from the inner sum,

$$\mathrm{Pf}(S) = \sum_{\ell-1}^{k-1} S_{k\ell}\frac{2}{2^m} \sum_{\substack{\mathbf{P}\in\mathcal{P}_m \\ k_1=k,\ell_1=\ell}} \varepsilon^{k\ell k_2\ell_2\ldots k_m\ell_m} S_{k_2\ell_2}\cdots S_{k_m\ell_m} + \sum_{\ell=k+1}^{n} S_{k\ell}\frac{2}{2^m} \sum_{\substack{\mathbf{P}\in\mathcal{P}_m \\ k_1=k,\ell_1=\ell}} \varepsilon^{k\ell k_2\ell_2\ldots k_m\ell_m} S_{k_2\ell_2}\cdots S_{k_m\ell_m}.$$

The following lemma implies that,

$$\mathrm{Pf}(S) = \sum_{\ell=1}^{k-1} S_{k\ell}(-1)^{k+\ell}\,\mathrm{Pf}(M_{k\ell}) + \sum_{\ell=k+1}^{n} S_{k\ell}(-1)^{k+\ell+1}\,\mathrm{Pf}(M_{k\ell}). \qquad \square$$

LEMMA B.3. *For all $1 \le k \ne \ell \le n$, let $M_{k\ell}$ be the matrix obtained from S by deleting rows k and ℓ and columns k and ℓ. Then*

$$\mathrm{Pf}(M_{k\ell}) = \mathrm{sgn}(k-\ell)\frac{(-1)^{k+\ell}}{2^{m-1}} \sum_{\substack{\mathbf{P}\in\mathcal{P}_m \\ k_1=k,\ell_1=\ell}} \varepsilon^{k\ell k_2\ell_2\ldots k_m\ell_m} S_{k_2\ell_2}\ldots S_{k_m\ell_m}.$$

PROOF. For all $1 \le k < \ell \le n$,

$$M_{k\ell} = (S_{i'j'}; 1 \le i,j \le n-2)$$

where, for each $1 \le i \le n-2$,

$$i' \begin{cases} i, & \text{if } 1 \le i \le k-1 \\ i+1, & \text{if } k \le i \le \ell-1 \\ i+2, & \text{if } \ell \le i \le n-2. \end{cases}$$

By definition,

$$\mathrm{Pf}(M_{k\ell}) = \frac{1}{2^{m-1}(m-1)!} \sum_{1\le i_1,\ldots,i_{n-2}\le n-2} \varepsilon^{i_1\ldots i_{n-2}} S_{i'_1 i'_2}\ldots S_{i'_{n-3}i'_{n-2}}.$$

We have

$$\varepsilon^{i_1\ldots i_{n-2}} = (-1)^{k+\ell-1}\varepsilon^{k\ell i'_1\ldots i'_{n-2}}$$

for all $1 \le i_1,\ldots,i_{n-2} \le n-2$. It follows that

$$\begin{aligned}\mathrm{Pf}(M_{k\ell}) &= \frac{(-1)^{k+\ell-1}}{2^{m-1}(m-1)!} \sum_{1\le i_1,\ldots,i_{n-2}\le n-2} \varepsilon^{k\ell i'_1\ldots i'_{n-2}} S_{i'_1 i'_2}\cdots S_{i'_{n-3}i'_{n-2}} \\ &= \frac{(-1)^{k+\ell-1}}{2^{m-1}(m-1)!} \sum_{1\le k_2,\ell_2\ldots,k_m,\ell_m\le n} \varepsilon^{k\ell k_2\ell_2\ldots k_m\ell_m} S_{k_2\ell_2}\cdots S_{k_m\ell_m}\end{aligned}$$

$$= \frac{(-1)^{k+\ell-1}}{2^{m-1}(m-1)!} \sum_{\substack{\mathbf{P}\in\mathcal{P}_m \\ k_1=k,\ell_1=\ell}} \varepsilon^{k\ell k_2\ell_2\ldots k_m\ell_m} S_{k_2\ell_2}\cdots S_{k_m\ell_m}.$$

If, on the other hand, $k > \ell$,

$$\begin{aligned}\operatorname{Pf}(M_{k\ell}) = \operatorname{Pf}(M_{\ell k}) &= \frac{(-1)^{k+\ell+1}}{2^{m-1}} \sum_{\substack{\mathbf{P}\in\mathcal{P}_m \\ k_1=\ell,\ell_1=k}} \varepsilon^{\ell k k_2\ell_2\ldots k_m\ell_m} S_{k_2\ell_2}\ldots S_{k_m\ell_m} \\ &= \frac{(-1)^{k+\ell}}{2^{m-1}} \sum_{\substack{\mathbf{P}\in\mathcal{P}_m \\ k_1=k,\ell_1=\ell}} \varepsilon^{k\ell k_2\ell_2\ldots k_m\ell_m} S_{k_2\ell_2}\ldots S_{k_m\ell_m}.\end{aligned}$$ □

PROPOSITION B.4. *Let*

$$S = \begin{pmatrix} \mathbf{0} & C \\ -C^t & \mathbf{0} \end{pmatrix}$$

where $C = (c_{ij})$ is a complex $m \times m$ matrix. Then,

$$\operatorname{Pf}(S) = (-1)^{m(m-1)/2}\det(C).$$

PROOF. The proof is by induction on $m \geq 1$. If $m = 1$, then, by Problem B.2,

$$\operatorname{Pf}\begin{pmatrix} 0 & S_{12} \\ S_{21} & 0 \end{pmatrix} = S_{12}.$$

Suppose $m > 1$. The matrix elements S_{ij}, $i, j = 1, \ldots, n = 2m$, of S are

$$S_{ij} = \begin{cases} 0, & \text{if } 1 \leq i, j \leq m \\ c_{ij-m}, & \text{if } 1 \leq i \leq m \text{ and } m+1 \leq j \leq n \\ -c_{ji-m}, & \text{if } m+1 \leq i \leq n \text{ and } 1 \leq j \leq m \\ 0, & \text{if } m+1 \leq i, j \leq n. \end{cases}$$

for each $k = 1, \ldots, m$, we have

$$M_{1k+m} = \begin{pmatrix} \mathbf{0} & C^{1,k} \\ -C^{1,k^t} & \mathbf{0} \end{pmatrix}.$$

where $C^{1,k}$ is the matrix of order $m-1$ obtained from the matrix C by deleting row 1 and column k. It now follows from Proposition B.2(b) and our induction hypothesis that

$$\begin{aligned}\operatorname{Pf}(S) &= \sum_{\ell=m+1}^{n} (-1)^\ell S_{1\ell} \operatorname{Pf}(M_{1\ell}) \\ &= \sum_{\ell=m+1}^{n} (-1)^\ell c_{1\ell-m} \operatorname{Pf}(M_{1\ell}) \\ &= \sum_{k=1}^{m} (-1)^{k+m} c_{1k} \operatorname{Pf}(M_{1k+m}) \\ &= \sum_{k=1}^{m} (-1)^{k+m} c_{1k} (-1)^{(m-1)(m-2)/2} \det(C^{1,k}).\end{aligned}$$

Collecting factors of minus one

$$\begin{aligned}\operatorname{Pf}(S) &= (-1)^{m+1}(-1)^{(m-1)(m-2)/2}\sum_{k=1}^{m}(-1)^{k+1}c_{1k}\det(C^{1,k})\\ &= (-1)^{m(m-1)/2}\sum_{k=1}^{m}(-1)^{k+1}c_{1k}\det(C^{1,k})\\ &= (-1)^{m(m-1)/2}\det(C).\end{aligned}$$

□

PROPOSITION B.5. *Let $S = (S_{i,j})$ be a skew symmetric matrix of even order $n = 2m$.*

(a) *For any matrix B of even order $n = 2m$,*

$$\operatorname{Pf}(B^tSB) = \det(B)\operatorname{Pf}(S).$$

(b) $\operatorname{Pf}(S)^2 = \det(S)$.

PROOF. (a) The (i_1, i_2) element of B^tSB is $\sum_{j_1,j_2} b_{j_1i_1}S_{j_1j_2}b_{j_2i_2}$. Hence

$$\begin{aligned}\operatorname{Pf}(B^tSB) &= \frac{1}{2^m m!}\sum_{i_1,\dots,i_n}\varepsilon^{i_1\dots i_n}\sum_{j_1,\dots,j_n} b_{j_1i_1}\cdots b_{j_ni_n}S_{j_1j_2}\cdots S_{j_{n-1}j_n}\\ &= \frac{1}{2^m m!}\sum_{j_1,\dots,j_n}\sum_{i_1,\dots,i_n}\varepsilon^{i_1\dots i_n}b_{j_1i_1}\cdots b_{j_ni_n}S_{j_1j_2}\cdots S_{j_{n-1}j_n}.\end{aligned}$$

The expression

$$\sum_{i_1,\dots,i_n}\varepsilon^{i_1\dots i_n}b_{j_1i_1}\cdots b_{j_ni_n}$$

is the determinant of the matrix whose ℓth row is the j_ℓth row of B. If any two of $j_1, \dots, j_n$ are equal, this determinant is zero. Otherwise it is $\varepsilon^{j_1\dots j_n}\det(B)$. Thus

$$\sum_{i_1,\dots,i_n}\varepsilon^{i_1\dots i_n}b_{j_1i_1}\cdots b_{j_ni_n} = \varepsilon^{j_1\dots j_n}\det(B)$$

and

$$\operatorname{Pf}(B^tSB) = \det(B)\frac{1}{2^m m!}\sum_{j_1,\dots,j_n}\varepsilon^{j_1\dots j_n}S_{j_1j_2}\cdots S_{j_{n-1}j_n} = \det(B)\operatorname{Pf}(S).$$

(b) It is enough to prove the identity for real, nonsingular matrices, since $\operatorname{Pf}(S)^2$ and $\det(S)$ are polynomials in the matrix elements S_{ij}, $i, j = 1, \dots, n$, of S. So, let S be real and nonsingular and, as in Lemma 1.10 (or Problem 1.9), let R be a real orthogonal matrix such that $R^tSR = T$ with

$$T = \bigoplus_{i=1}^{m}\begin{bmatrix}0 & \alpha_i\\ -\alpha_i & 0\end{bmatrix}$$

for some real numbers $\alpha_1, \dots, \alpha_m$. Then, $\det R = \pm 1$, so that, by part (a) and Problem B.3,

$$\operatorname{Pf}(S) = \pm\det(R)\operatorname{Pf}(S) = \pm\operatorname{Pf}(R^tSR) = \pm\operatorname{Pf}(T) = \pm\alpha_1\cdots\alpha_m$$

and

$$\det(S) = \det(R^tSR) = \det T = \alpha_1^2\dots\alpha_m^2 = \operatorname{Pf}(S)^2.$$

□

PROBLEM B.4. Let S be a skew symmetric $D \times D$ matrix with D odd. Prove that $\det S = 0$.

APPENDIX C

Propagator Bounds

The propagator, or covariance, for many-fermion models is the Fourier transform of

$$C_{\sigma,\sigma'}(k) = \frac{\delta_{\sigma,\sigma'}}{\imath k_0 - e(\mathbf{k})}$$

where $k = (k_0, \mathbf{k})$ and $e(\mathbf{k})$ is the one particle dispersion relation minus the chemical potential. For this appendix, the spins σ, σ' play no role, so we suppress them completely. We also restrict our attention to two space dimensions (i.e., $\mathbf{k} \in \mathbb{R}^2$, $k \in \mathbb{R}^3$) though it is trivial to extend the results of this appendix to any number of space dimensions. We assume that $e(\mathbf{k})$ is a C^4 (though $C^{2+\varepsilon}$ would suffice) function that has a nonempty, compact zero set $\mathcal{F}$, called the Fermi curve. We further assume that $\nabla e(\mathbf{k})$ does not vanish for $\mathbf{k} \in \mathcal{F}$, so that $\mathcal{F}$ is itself a reasonable smooth curve. At low temperatures only those momenta with $k_0 \approx 0$ and $\mathbf{k}$ near $\mathcal{F}$ are important, so we replace the above propagator with

$$C(k) = \frac{U(k)}{\imath k_0 - e(\mathbf{k})}.$$

The precise ultraviolet cutoff, $U(k)$, shall be chosen shortly. It is a C_0^∞ function which takes values in $[0,1]$, is identically 1 for $k_0^2 + e(\mathbf{k})^2 \le 1$ and vanishes for $k_0^2 + e(\mathbf{k})^2$ larger than some constant.

We slice momentum space into shells around the Fermi surface. To do this, we fix $M > 1$ and choose a function $\nu \in C_0^\infty([M^{-2}, M^2])$ that takes values in $[0,1]$, is identically 1 on $[M^{-1/2}, M^{1/2}]$ and obeys

$$\sum_{j=0}^{\infty} \nu(M^{2j}x) = 1$$

for $0 < x < 1$. The jth shell is defined to be the support of

$$\nu^{(j)}(k) = \nu\Big(M^{2j}\left(k_0^2 + e(\mathbf{k})^2\right)\Big).$$

By construction, the jth shell is a subset of

$$\left\{k \;\middle|\; \frac{1}{M^{j+1}} \le |\imath k_0 - e(\mathbf{k})| \le \frac{1}{M^{j-1}}\right\}.$$

Setting

$$C^{(j)}(k) = C(k)\nu^{(j)}(k)$$

and $U(k) = \sum_{j=0}^{\infty} \nu^{(j)}(k)$ we have

$$C(k) = \sum_{j=0}^{\infty} C^{(j)}(k).$$

Given any function $\chi(\mathbf{k}')$ on the Fermi curve $\mathcal{F}$, we define

$$C_{\chi}^{(j)}(k) = C^{(j)}(k)\chi\big(\mathbf{k}'(k)\big)$$

where, for $k = (k_0, \mathbf{k})$, $\mathbf{k}'(k)$ is any reasonable "projection" of $\mathbf{k}$ onto the Fermi curve. In the event that $\mathcal{F}$ is a circle of radius $k_{\mathcal{F}}$ centered on the origin, it is natural to choose $\mathbf{k}'(k) = (k_{\mathcal{F}}/|\mathbf{k}|)\mathbf{k}$. For general $\mathcal{F}$, one can always construct, in a tubular neighborhood of $\mathcal{F}$, a C^∞ vector field that is transverse to $\mathcal{F}$, and then define $\mathbf{k}'(k)$ to be the unique point of $\mathcal{F}$ that is on the same integral curve of the vector field as $\mathbf{k}$ is. See [**FST**, Lemma 2.1].

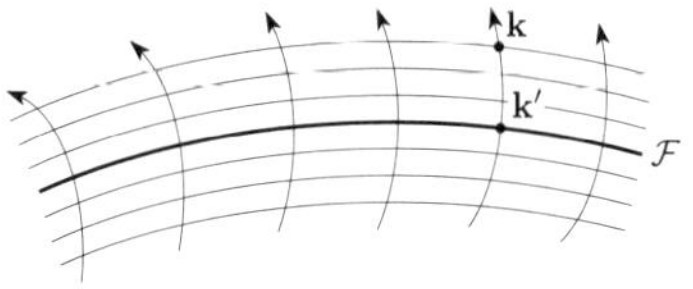

To analyze the Fourier transform of $C^{(j)}(k)$, we further decompose the jth shell into more or less rectangular "sectors." To do so, we fix $\mathfrak{l}_j \in [1/M_j, 1/M^{j/2}]$ and choose a partition of unity

$$1 = \sum_{s \in \Sigma^{(j)}} \chi_s^{(j)}(\mathbf{k}')$$

of the Fermi curve $\mathcal{F}$ with each $\chi_s^{(j)}$ supported on an interval of length $\mathfrak{l}_j$ and obeying

$$\sup_{\mathbf{k}'} |\partial_{\mathbf{k}'}^m \chi_s^{(j)}| \le \frac{\text{const}}{\mathfrak{l}_j^m} \qquad \text{for } m \le 4.$$

Here $\partial_{\mathbf{k}'}$ is the derivative along the Fermi curve. If $\mathbf{k}'(t)$ is a parametrization of the Fermi curve by arc length, with the standard orientation, then $\partial_{\mathbf{k}'} f(\mathbf{k}')|_{\mathbf{k}'=\mathbf{k}'(t)} = df\big(\mathbf{k}'(t)\big)/dt$.

PROPOSITION C.1. *Let $\chi(\mathbf{k}')$ be a C^4 function on the Fermi curve $\mathcal{F}$ which takes values in $[0,1]$, which is supported on an interval of length $\mathfrak{l}_j \in [1/M^j, 1/M^{j/2}]$ and whose derivatives obey*

$$\sup_{\mathbf{k}'} |\partial_{\mathbf{k}'}^n \chi(\mathbf{k}')| \le \frac{K_\chi}{\mathfrak{l}_j^n} \qquad \textit{for } n \le 4.$$

Fix any point $\mathbf{k}'_c$ in the support of χ. Let $\hat{\mathbf{t}}$ and $\hat{\mathbf{n}}$ be unit tangent and normal vectors to the Fermi curve at $\mathbf{k}'_c$ and set

$$\rho(x,y) = 1 + M^{-j}|x_0 - y_0| + M^{-j}|(\mathbf{x}-\mathbf{y})\cdot\hat{\mathbf{n}}| + \mathfrak{l}_j|(\mathbf{x}-\mathbf{y})\cdot\hat{\mathbf{t}}|.$$

Let ϕ be a C_0^4 function which takes values in $[0,1]$ and set $\phi^{(j)} = \phi\big(M^{2j}[k_0^2 + e(\mathbf{k})^2]\big)$. For any function $W(k)$ define

$$W_{\chi,\phi}^{(j)}(x,y) = \int \frac{d^3k}{(2\pi)^3} e^{\imath k\cdot(x-y)} W(k)\phi^{(j)}(k)\chi\big(\mathbf{k}'(k)\big).$$

There is a constant, const, *depending on K_χ, ϕ and $e(\mathbf{k})$, but independent of M, j, x and y such that*

$$|W^{(j)}_{\chi,\phi}(x,y)| \le \text{const}\,\frac{\mathfrak{l}_j}{M^{2j}}\rho(x,y)^{-4} \\ \times \max_{\substack{\alpha\in\mathbb{N}^3\\ |\alpha|\le 4}} \sup_{k\in\operatorname{supp}\chi\phi^{(j)}} \frac{\mathfrak{l}_j^{\alpha_2}}{M^{j(\alpha_0+\alpha_1)}}|\partial_{k_0}^{\alpha_0}(\hat{\mathbf{n}}\cdot\nabla_{\mathbf{k}})^{\alpha_1}(\hat{\mathbf{t}}\cdot\nabla_{\mathbf{k}})^{\alpha_2}W(k)|$$

where $\alpha = (\alpha_0, \alpha_1, \alpha_2)$ and $|\alpha| = \alpha_0 + \alpha_1 + \alpha_2$.

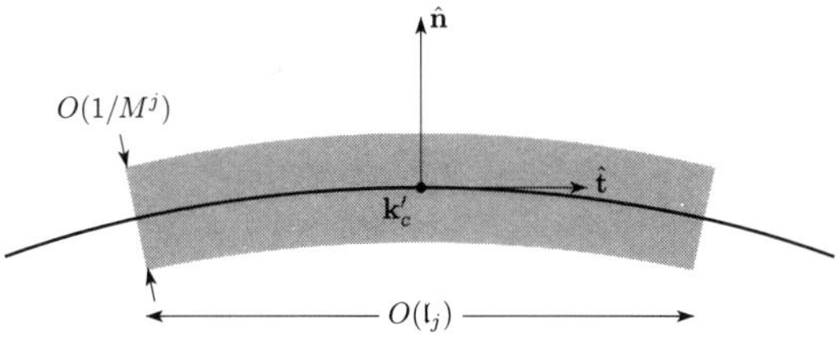

PROOF. Use S to denote the support of $\phi^{(j)}(k)\chi\big(\mathbf{k}'(k)\big)$. If $\phi(x)$ vanishes for $x \ge Q^2$, then $\phi^{(j)}(k)$ vanishes unless $|k_0| \le QM^j$ and $|e(\mathbf{k})| \le QM^j$. If $k \in S$, then k_0 lies in an interval of length $2QM^{-j}$, the component of $\mathbf{k}$ tangential to $\mathcal{F}$ lies in an interval of length const $\mathfrak{l}_j$ and, by Problem C.1 below, the component of $\mathbf{k}$ normal to $\mathcal{F}$ lies in an interval of length $2C'QM^{-j}$. Thus S has volume at most const $M^{-2j}\mathfrak{l}_j$ and

$$\sup_{x,y}|W^{(j)}_{\chi,\phi}(x,y)| \le \operatorname{vol}(S)\sup_{k\in S}|W(k)| \le \text{const}\,\frac{\mathfrak{l}_j}{M^{2j}}\sup_{k\in S}|W(k)|.$$

We define ρ as a sum of four terms. Multiplying out ρ^4 gives us a sum of 4^4 terms, each of the form $|(x_0-y_0)/M^j|^{\beta_0}|(\mathbf{x}-\mathbf{y})\cdot\hat{\mathbf{n}}/M^j|^{\beta_1}|\mathfrak{l}_j(\mathbf{x}-\mathbf{y})\cdot\hat{\mathbf{t}}|^{\beta_2}$ with $|\beta| \le 4$. To bound $\sup_{x,y}\rho(x,y)^4|W^{(j)}_{\chi,\phi}(x,y)|$ by 4^4C, it suffices to bound

$$\left|\left(\frac{x_0-y_0}{M^j}\right)^{\beta_0}\left(\frac{(x-y)\cdot\hat{\mathbf{n}}}{M^j}\right)^{\beta_1}\big(\mathfrak{l}_j(\mathbf{x}-\mathbf{y})\cdot\hat{\mathbf{t}}\big)^{\beta_2}W^{(j)}_{\chi,\phi}(x,y)\right| \\ = \left|\int\frac{d^3k}{(2\pi)^3}e^{ik\cdot(x-y)}\left(\frac{1}{M^j}\partial_{k_0}\right)^{\beta_0}\left(\frac{1}{M^j}\hat{\mathbf{n}}\cdot\nabla_{\mathbf{k}}\right)^{\beta_1}(\mathfrak{l}_j\hat{\mathbf{t}}\cdot\nabla_{\mathbf{k}})^{\beta_2}\Big(W(k)\phi^{(j)}(k)\chi\big(\mathbf{k}'(k)\big)\Big)\right|$$

by C for all x, $y \in \mathbb{R}^3$ and $\beta \in \mathbb{N}^3$ with $|\beta| \le 4$. The volume of the domain of integration is still bounded by const $\mathfrak{l}_j/M^{2j}$, so by the product rule, to prove the desired bound it suffices to prove that

$$\max_{|\beta|\le 4}\sup_{k\in S}\left|\left(\frac{1}{M^j}\partial_{k_0}\right)^{\beta_0}\left(\frac{1}{M^j}\hat{\mathbf{n}}\cdot\nabla_{\mathbf{k}}\right)^{\beta_1}(\mathfrak{l}_j\hat{\mathbf{t}}\cdot\nabla_{\mathbf{k}})^{\beta_2}\Big(\phi^{(j)}(k)\chi\big(\mathbf{k}'(k)\big)\Big)\right| \le \text{const}.$$

Since $\mathfrak{l}_j \ge 1/M^j$ and all derivatives of $\mathbf{k}'(k)$ to order 4 are bounded,

$$\max_{|\beta|\le 4}\sup_{k}\left|\left(\frac{1}{M^j}\partial_{k_0}\right)^{\beta_0}\left(\frac{1}{M^j}\hat{\mathbf{n}}\cdot\nabla_{\mathbf{k}}\right)^{\beta_1}(\mathfrak{l}_j\hat{\mathbf{t}}\cdot\nabla_{\mathbf{k}})^{\beta_2}\chi\big(\mathbf{k}'(k)\big)\right| \\ \le \text{const}\max_{\beta_1+\beta_2\le 4}\frac{\mathfrak{l}_j^{\beta_2}}{M^{j\beta_1}}\frac{1}{\mathfrak{l}_j^{\beta_1+\beta_2}} \le \text{const}$$

so, by the product rule, it suffices to prove

$$\max_{|\beta|\le 4}\sup_{k\in S}\left|\left(\frac{1}{M^j}\partial_{k_0}\right)^{\beta_0}\left(\frac{1}{M^j}\hat{\mathbf{n}}\cdot\nabla_{\mathbf{k}}\right)^{\beta_1}(\mathfrak{l}_j\hat{\mathbf{t}}\cdot\nabla_{\mathbf{k}})^{\beta_2}\phi^{(j)}(k)\right|\le\text{const.}$$

Set $I=\{1,\ldots,|\beta|\}$,

$$d_i=\begin{cases}\frac{1}{M^j}\partial_{k_0}, & \text{if } 1\le i\le\beta_0\\ \frac{1}{M^j}\hat{\mathbf{n}}\cdot\nabla_{\mathbf{k}}, & \text{if } \beta_0+1\le i\le\beta_0+\beta_1\\ \mathfrak{l}_j\hat{\mathbf{t}}\cdot\nabla_{\mathbf{k}}, & \text{if } \beta_0+\beta_1+1\le i\le|\beta|\end{cases}$$

and, for each $I'\subset I$, $d^{I'}=\prod_{i\in I'}d_i$. By Problem C.2, below,

$$d^I\phi^{(j)}(k)=\sum_{m=1}^{|\beta|}\sum_{(I_1,\ldots,I_m)\in\mathcal{P}_m}\frac{d^m\phi}{dx^m}\Big(M^{2j}\big(k_0^2+e(\mathbf{k})^2\big)\Big)\prod_{i=1}^m M^{2j}d^{I_i}\big(k_0^2+e(\mathbf{k})^2\big)$$

where $\mathcal{P}_m$ is the set of all partitions of I into m nonempty subsets $I_1,\ldots,I_m$ with, for all $i<i'$, the smallest element of I_i smaller than the smallest element of $I_{i'}$. For all $m\le 4$, $\left|d^m\phi/dx^m\big(M^{2j}\big(k_0^2+e(\mathbf{k})^2\big)\big)\right|$ is bounded by a constant independent of j, so to prove the proposition, it suffices to prove that

$$\max_{|\beta|\le 4}\sup_{k\in S}\left|M^{2j}\left(\frac{1}{M^j}\partial_{k_0}\right)^{\beta_0}\left(\frac{1}{M^j}\hat{\mathbf{n}}\cdot\nabla_{\mathbf{k}}\right)^{\beta_1}(\mathfrak{l}_j\hat{\mathbf{t}}\cdot\nabla_{\mathbf{k}})^{\beta_2}(k_0^2+e(\mathbf{k})^2)\right|\le\text{const.}$$

If $\beta_0\ne 0$

$$M^{2j}\left(\frac{1}{M^j}\partial_{k_0}\right)^{\beta_0}\left(\frac{1}{M^j}\hat{\mathbf{n}}\cdot\nabla_{\mathbf{k}}\right)^{\beta_1}(\mathfrak{l}_j\hat{\mathbf{t}}\cdot\nabla_{\mathbf{k}})^{\beta_2}\big(k_0^2+e(\mathbf{k})^2\big)$$
$$=\begin{cases}2k_0M^j, & \text{if } \beta_0=1,\ \beta_1=\beta_2=0\\ 2, & \text{if } \beta_0=2,\ \beta_1=\beta_2=0\\ 0, & \text{otherwise}\end{cases}$$

is bounded, independent of j, since $|k_0|\le\text{const}\,1/M^j$ on S. Thus it suffices to consider $\beta_0=0$. Applying the product rule once again, this time to the derivatives acting on $M^{2j}e(\mathbf{k})^2=[M^je(\mathbf{k})][M^je(\mathbf{k})]$, it suffices to prove

$$\max_{|\beta|\le 4}\sup_{k\in S}\left|M^j\left(\frac{1}{M^j}\hat{\mathbf{n}}\cdot\nabla_{\mathbf{k}}\right)^{\beta_1}(\mathfrak{l}_j\hat{\mathbf{t}}\cdot\nabla_{\mathbf{k}})^{\beta_2}e(\mathbf{k})\right|\le\text{const.}\tag{C.1}$$

If $\beta_1=\beta_2=0$, this follows from the fact that $|e(\mathbf{k})|\le\text{const}\,1/M^j$ on S. If $\beta_1\ge 1$ or $\beta_2\ge 2$, it follows from $M^j\mathfrak{l}_j^{\beta_2}/M^{\beta_1 j}\le 1$. (Recall that $\mathfrak{l}_j\le 1/M_{j/2}$.) This leaves only $\beta_1=0$, $\beta_2=1$. If $\hat{\mathbf{t}}\cdot\nabla_{\mathbf{k}}e(\mathbf{k})$ is evaluated at $\mathbf{k}=\mathbf{k}_c'$, it vanishes, since $\nabla_{\mathbf{k}}e(\mathbf{k}_c')$ is parallel to $\hat{\mathbf{n}}$. The second derivative of e is bounded so that,

$$\begin{aligned}M^j\mathfrak{l}_j\sup_{k\in S}|\hat{\mathbf{t}}\cdot\nabla_{\mathbf{k}}e(\mathbf{k})|&=M^j\mathfrak{l}_j\sup_{k\in S}|\hat{\mathbf{t}}\cdot\nabla_{\mathbf{k}}e(\mathbf{k})-\hat{\mathbf{t}}\cdot\nabla_{\mathbf{k}}e(\mathbf{k}_c')|\\&\le\text{const}\,M^j\mathfrak{l}_j\sup_{k\in S}|\mathbf{k}-\mathbf{k}_c'|\\&\le\text{const}\,M^j\mathfrak{l}_j^2\le\text{const}\end{aligned}$$

since $\mathfrak{l}_j\le 1/M^{j/2}$. □

PROBLEM C.1. Prove, under the hypotheses of Proposition C.1, that there are constants C, $C' > 0$, such that if $|\mathbf{k} - \mathbf{k}_c'| \le C$, then there is a point $\mathbf{p}' \in \mathcal{F}$ with $(\mathbf{k} - \mathbf{p}') \cdot \hat{\mathbf{t}} = 0$ and $|\mathbf{k} - \mathbf{p}'| \le C'|e(\mathbf{k})|$.

PROBLEM C.2. Let $f\colon \mathbb{R}^d \to \mathbb{R}$ and $g\colon \mathbb{R} \to \mathbb{R}$. Let $1 \le i_1, \dots, i_n \le d$. Prove that

$$\Big(\prod_{\ell=1}^{n} \frac{\partial}{\partial x_{i_\ell}}\Big) g(f(x)) = \sum_{m=1}^{n} \sum_{(I_1,\dots,I_m)\in\mathcal{P}_m^{(n)}} g^{(m)}(f(x)) \prod_{p=1}^{m} \prod_{\ell\in I_p} \frac{\partial}{\partial x_{i_\ell}} f(x)$$

where $\mathcal{P}_m^{(n)}$ is the set of all partitions of $(1, \dots, n)$ into m nonempty subsets $I_1, \dots, I_m$ with, for all $i < i'$, the smallest element of I_i smaller than the smallest element of $I_{i'}$.

COROLLARY C.2. *Under the hypotheses of Proposition* C.1,

$$\sup |W_{\chi,\phi}^{(j)}(x, y)| \le \text{const}\, \frac{\mathfrak{l}_j}{M^{2j}} \sup_{k\in \operatorname{supp} \chi\phi^{(j)}} |W(k)|$$

and

$$\begin{aligned}\sup_x \int dy |W_{\chi,\phi}^{(j)}(x, y)|, &\sup_y \int dx |\, W_{\chi,\phi}^{(j)}(x, y)| \\ &\le \text{const} \max_{\substack{\alpha\in\mathbb{N}^3\\ |\alpha|\le 4}} \sup_{k\in \operatorname{supp} \chi\phi^{(j)}} \frac{\mathfrak{l}_j^{\alpha_2}}{M^{j(\alpha_0+\alpha_1)}} |\partial_{k_0}^{\alpha_0} (\hat{\mathbf{n}} \cdot \nabla_{\mathbf{k}})^{\alpha_1} (\hat{\mathbf{t}} \cdot \nabla_{\mathbf{k}})^{\alpha_2} W(k)|.\end{aligned}$$

PROOF. The first claim is an immediate consequence of Proposition C.1 since $\rho \ge 1$. For the second statement, use

$$\sup_x \int dy \frac{1}{\rho(x, y)^4}, \sup_y \int dx \frac{1}{\rho(x, y)^4} = \sup_y \int dx \frac{1}{\rho(x - y, 0)^4} = \int dx' \frac{1}{\rho(x', 0)^4}$$

with $x' = x - y$. Subbing in the definition of ρ,

$$\begin{aligned}\int dx \frac{1}{\rho(x, 0)^4} &= \int dx\, \frac{1}{[1 + M^{-j}|x_0| + M^{-j}|\mathbf{x}\cdot\hat{\mathbf{n}}| + \mathfrak{l}_j|\mathbf{x}\cdot\hat{\mathbf{t}}|]^4} \\ &= M^{2j} \frac{1}{\mathfrak{l}_j} \int dz\, \frac{1}{[1 + |z_0| + |z_1| + |z_2|]^4} \\ &\le \text{const}\, M^{2j} \frac{1}{\mathfrak{l}_j}.\end{aligned}$$

We made the change of variables $x_0 = M^j z_1$, $\mathbf{x}\cdot\hat{\mathbf{n}} = M^j z_1$, $\mathbf{x}\cdot\hat{\mathbf{t}} = z_2/\mathfrak{l}_j$. □

COROLLARY C.3. *Under the hypotheses of Proposition* C.1,

$$\sup |C_\chi^{(j)}(x, y)| \le \text{const}\, \frac{\mathfrak{l}_j}{M^j}$$

and

$$\sup_x \int dy\, C_\chi^{(j)}(x, y)|, \sup_y \int dx\, |C_\chi^{(j)}(x, y)| \le \text{const}\, M^j.$$

PROOF. Apply Corollary C.2 with $W(k) = 1/(\imath k_0 - e(\mathbf{k}))$ and $\phi = \nu$. To achieve the desired bounds, we need

$$\begin{aligned}\max_{|\alpha|\le 4} \sup_{k\in \operatorname{supp} \chi\nu^{(j)}} \left| \Big(\frac{1}{M^j}\partial_{k_0}\Big)^{\alpha_0} \Big(\frac{1}{M^j}\hat{\mathbf{n}}\cdot\nabla_{\mathbf{k}}\Big)^{\alpha_1} (\mathfrak{l}_j\hat{\mathbf{t}}\cdot\nabla_{\mathbf{k}})^{\alpha_2} \frac{1}{\imath k_0 - e(\mathbf{k})} \right| \\ \le \text{const}\, M^j.\end{aligned}$$

In the notation of the proof of Proposition C.1, with β replaced by α,

$$\begin{aligned} d^I \nu^{(j)}(k) &= \sum_{m=1}^{|\alpha|} (-1)^m m! \sum_{(I_1,\dots,I_m)\in\mathcal{P}_m} \left(\frac{1}{\imath k_0 - e(\mathbf{k})}\right)^{m+1} \prod_{i=1}^{m} d^{I_i}\big(\imath k_0 - e(\mathbf{k})\big) \\ &= M^j \sum_{m=1}^{|\alpha|} (-1)^m m! \sum_{(I_1,\dots,I_m)\in\mathcal{P}_m} \left(\frac{1/M^j}{\imath k_0 - e(\mathbf{k})}\right)^{m+1} \prod_{i=1}^{m} M^j d^{I_i}\big(\imath k_0 - e(\mathbf{k})\big). \end{aligned}$$

On the support of $\chi\nu^{(j)}$, $|\imath k_0 - e(\mathbf{k})| \ge \text{const}\, 1/M^j$ so that $\big((1/M^j)/\big(\imath k_0 - e(\mathbf{k})\big)\big)^{m+1}$ is bounded uniformly in j. That $M^j d^{I_i}\big(\imath k_0 - e(\mathbf{k})\big)$ is bounded uniformly in j is immediate from (C.1),

$$M^j \left(\frac{1}{M^j}\partial_{k_0}\right)^{\beta_0} \left(\frac{1}{M^j}\hat{\mathbf{n}}\cdot\nabla_{\mathbf{k}}\right)^{\beta_1} (\mathfrak{l}_j \hat{\mathbf{t}}\cdot\nabla_{\mathbf{k}})^{\beta_2}(\imath k_0)$$
$$= \begin{cases} \imath k_0 M^j, & \text{if } \beta_0 = \beta_1 = \beta_2 = 0 \\ \imath, & \text{if } \beta_0 = 1,\ \beta_1 = \beta_2 = 0 \\ 0, & \text{otherwise} \end{cases}$$

and the fact that $|k_0| \le \text{const}\, 1/M^j$ on the support of $\chi\nu^{(j)}$. □

APPENDIX D

Problem Solutions

Chapter 1. Fermionic Functional Integrals

PROBLEM 1.1. *Let $\mathcal{V}$ be a complex vector space of dimension D. Let $s \in \bigwedge \mathcal{V}$. Then s has a unique decomposition $s = s_0 + s_1$ with $s_0 \in \mathbb{C}$ and $s_1 \in \bigoplus_{n=1}^{D} \bigwedge^n \mathcal{V}$. Prove that, if $s_0 \neq 0$, then there is a unique $s' \in \bigwedge \mathcal{V}$ with $ss' = 1$ and a unique $s'' \in \bigwedge \mathcal{V}$ with $s''s = 1$ and furthermore*

$$s' = s'' = \frac{1}{s_0} + \sum_{n=1}^{D} (-1)^n \frac{s_1^n}{s_0^{n+1}}.$$

SOLUTION. Note that if $n > D$, then $s_1^n = 0$. Define

$$\tilde{s} = \frac{1}{s_0} + \sum_{n=1}^{D} (-1)^n \frac{s_1^n}{s_0^{n+1}}.$$

Then

$$\begin{aligned}
\tilde{s}s = s\tilde{s} &= (s_0 + s_1)\left[\frac{1}{s_0} + \sum_{n=1}^{D}(-1)^n \frac{s_1^n}{s_0^{n+1}}\right] \\
&= \left[1 + \sum_{n=1}^{D}(-1)^n \frac{s_1^n}{s_0^n}\right] + \left[\frac{s_1}{s_0} + \sum_{n=1}^{D}(-1)^n \frac{s_1^{n+1}}{s_0^{n+1}}\right] \\
&= \left[1 + \sum_{n=1}^{D}(-1)^n \frac{s_1^n}{s_0^n}\right] + \left[\sum_{n-0}^{D-1}(-1)^n \frac{s_1^{n+1}}{s_0^{n+1}}\right] \\
&= \left[1 + \sum_{n=1}^{D}(-1)^n \frac{s_1^n}{s_0^n}\right] + \left[\sum_{n'=1}^{D}(-1)^{n'-1} \frac{s^{n'}}{s_0^{n'}}\right] = 1.
\end{aligned}$$

Thus $\tilde{s}$ satisfies $\tilde{s}s = s\tilde{s} = 1$. If some other s' satisfies $ss' = 1$, then $s' = 1s' = \tilde{s}ss' = \tilde{s}1 = \tilde{s}$ and if some other s'' satisfies $s''s = 1$, then $s'' = s''1 = s''s\tilde{s} = 1\tilde{s} = \tilde{s}$. □

PROBLEM 1.2. *Let $\mathcal{V}$ be a complex vector space of dimension D. Every element s of $\bigwedge \mathcal{V}$ has a unique decomposition $s = s_0 + s_1$ with $s_0 \in \mathbb{C}$ and $s_1 \in \bigoplus_{n=1}^{D} \bigwedge^n \mathcal{V}$. Define*

$$e^s = e^{s_0}\left\{\sum_{n=0}^{D} \frac{1}{n!} s_1^n\right\}.$$

Prove that if $s,\ t \in \bigwedge \mathcal{V}$ with $st = ts$, then, for all $n \in \mathbb{N}$,

$$(s+t)^n = \sum_{m=0}^{n} \binom{n}{m} s^m t^{n-m}$$

and

$$e^s e^t = e^t e^s = e^{s+t}.$$

SOLUTION. The proof that $(s+t)^n = \sum_{m=0}^n \binom{n}{m} s^m t^{t-m}$ is by induction on n. The case $n = 1$ is obvious. Suppose that $(s+t)^n = \sum_{m=0}^n \binom{n}{m} s^m t^{n-m}$ for some n. Since $st = ts$, we have $ss^m t^n = s^m t^n s$ for all nonnegative integers m and n, so that

$$\begin{aligned}
(s+t)^{n+1} &= s(s+t)^n + (s+t)^n t \\
&= \sum_{m=0}^{n} \binom{n}{m} s^{m+1} t^{n-m} + \sum_{m=0}^{n} \binom{n}{m} s^m t^{n+1-m} \\
&= \sum_{m=1}^{n+1} \binom{n}{m-1} s^m t^{n+1-m} + \sum_{m=0}^{n} \binom{n}{m} s^m t^{n+1-m} \\
&= s^{n+1} + \sum_{m=1}^{n} \binom{n}{m-1} s^m t^{n+1-m} + \sum_{m=1}^{n} \binom{n}{m} s^m t^{n+1-m} + t^{n+1} \\
&= t^{n+1} + \sum_{m=1}^{n} \left[\frac{n!}{(m-1)!(n-m+1)!} + \frac{n!}{m!(n-m)!} \right] s^m t^{n+1-m} + s^{n+1} \\
&= t^{n+1} + \sum_{m=1}^{n} \binom{n+1}{m} \left[\frac{m}{n+1} + \frac{n+1-m}{n+1} \right] s^m t^{n+1-m} + s^{n+1} \\
&= \sum_{m=0}^{n+1} \binom{n+1}{m} s^m t^{n+1-m}.
\end{aligned}$$

For the second part, since s_0, t_0, $s_1 = s - s_0$ and $t_1 = t - t_0$ all commute with each other, and $s_1^n = t_1^n = 0$ for all $n > D$

$$\begin{aligned}
e^s e^t = e^t e^s = e^{s_0} \left\{ \sum_{m=0}^{\infty} \frac{1}{m!} s_1^m \right\} e^{t_0} \{ \sum_{n=0}^{\infty} \frac{1}{n!} t_1^n \} &= e^{s_0 t^{t_0}} \left\{ \sum_{m,n=0}^{\infty} \frac{1}{m!n!} s_1^m t_1^n \right\} \\
&= e^{s_0} e^{t_0} \left\{ \sum_{N=0}^{\infty} \sum_{m=0}^{N} \frac{1}{m!(N-m)!} s_1^m t_1^{N-m} \right\} \quad \text{where } N = n+m \\
&= e^{s_0} e^{t_0} \left\{ \sum_{N=0}^{\infty} \frac{1}{N!} \sum_{m=0}^{N} \binom{N}{m} s_1^m t_1^{N-m} \right\} \\
&= e^{s_0 t_0} \left\{ \sum_{N=0}^{\infty} \frac{1}{N!} (s_1 + t_1)^N \right\} = e^{s+t}.
\end{aligned}$$

□

PROBLEM 1.3. *Use the notation of Problem 1.2. Let, for each $\alpha \in \mathbb{R}$, $s(\alpha) \in \bigwedge \mathcal{V}$. Assume that $s(\alpha)$ is differentiable with respect to α and that $s(\alpha)s(\beta) = s(\beta)s(\alpha)$ for all α and β. Prove that*

$$\frac{d}{d\alpha} s(\alpha)^n = n s(\alpha)^{n-1} \frac{d}{d\alpha} s(\alpha)$$

and

$$\frac{d}{d\alpha} e^{s(\alpha)} = e^{s(\alpha)} \frac{d}{d\alpha} s(\alpha).$$

SOLUTION. First, let for each $\alpha \in \mathbb{R}$, $t(\alpha) \in \bigwedge \mathcal{V}$ and assume that $t(\alpha)$ is differentiable with respect to α. Taking the limit of

$$\frac{s(\alpha+h)t(\alpha+h)-s(\alpha)t(\alpha)}{h} = \frac{s(\alpha+h)-s(\alpha)}{h}t(\alpha+h) + s(\alpha)\frac{t(\alpha+h)-t(\alpha)}{h}$$

a h tends to zero gives that $s(\alpha)t(\alpha)$ is differentiable and obeys the usual product rule

$$\frac{d}{d\alpha}[s(\alpha)t(\alpha)] = s'(\alpha)t(\alpha) + s(\alpha)t'(\alpha)$$

Differentiating $s(\alpha)s(\beta) = s(\beta)s(\alpha)$ with respect to α gives that $s'(\alpha)s(\beta) = s(\beta)s'(\alpha)$ for all α and β too. If for some n, $ds(\alpha)^n/d\alpha = ns(\alpha)^{n-1}s'(\alpha)$, then

$$\begin{aligned}\frac{d}{d\alpha}s(\alpha)s(\alpha)^n &= s'(\alpha)s(\alpha)^n + s(\alpha)\frac{d}{d\alpha}s(\alpha)^n \\ &= s'(\alpha)s(\alpha)^n + ns(\alpha)s(\alpha)^{n-1}s'(\alpha) \\ &= (n+1)s(\alpha)^n s'(\alpha)\end{aligned}$$

and the first part follows easily by induction. For the second part, write $s(\alpha) = s_0(\alpha) + s_1(\alpha)$ with $s_0(\alpha) \in \mathbb{C}$ and $s_1(\alpha) \in \bigoplus_{n=1}^{D} \bigwedge^n \mathcal{V}$. Then $s_0(\alpha)$ and $s_1(\beta)$ commute for all α and β and

$$\begin{aligned}\frac{d}{d\alpha}e^{s(\alpha)} &= \frac{d}{d\alpha}\left[e^{s_0(\alpha)}\left\{\sum_{n=0}^{D}\frac{1}{n!}s_1(\alpha)^n\right\}\right] \\ &= \frac{d}{d\alpha}\left[e^{s_0(\alpha)}\left\{\sum_{n=0}^{D+1}\frac{1}{n!}s_1(\alpha)^n\right\}\right] \\ &= s_0'(\alpha)e^{s_0(\alpha)}\left\{\sum_{n=0}^{D+1}\frac{1}{n!}s_1(\alpha)^n\right\} + e^{s_0(\alpha)}\left\{\sum_{n=1}^{D+1}\frac{1}{(n-1)!}s_1(\alpha)^{n-1}\right\}s_1'(\alpha) \\ &= s_0'(\alpha)e^{s_0(\alpha)}\left\{\sum_{n=0}^{D}\frac{1}{n!}s_1(\alpha)^n\right\} + e^{s_0(\alpha)}\left\{\sum_{n=0}^{D}\frac{1}{n!}s_1(\alpha)^n\right\}s_1'(\alpha) \\ &= e^{s_0(\alpha)}\left\{\sum_{n=0}^{D}\frac{1}{n!}s_1(\alpha)^n\right\}\{s_0'(\alpha)+s_1'(\alpha)\} \\ &= e^{s(\alpha)}s'(\alpha).\end{aligned}$$

□

PROBLEM 1.4. *Use the notation of Problem 1.2. If $s_0 > 0$, define*

$$\ln s = \ln s_0 + \sum_{n=1}^{D}\frac{(-1)^{n-1}}{n}\left(\frac{s_1}{s_0}\right)^n$$

with $\ln s_0 \in \mathbb{R}$.

(a) *Let, for each $\alpha \in \mathbb{R}$, $s(\alpha) \in \bigwedge \mathcal{V}$. Assume that $s(\alpha)$ is differentiable with respect to α, that $s(\alpha)s(\beta) = s(\beta)s(\alpha)$ for all α and β and that $s_0(\alpha) > 0$ for all α. Prove that*

$$\frac{d}{d\alpha}\ln s(\alpha) = \frac{s'(\alpha)}{s(\alpha)}.$$

(b) *Prove that if $s \in \bigwedge \mathcal{V}$ with $s_0 \in \mathbb{R}$, then*

$$\ln e^s = s.$$

Prove that if $s \in \bigwedge \mathcal{V}$ *with* $s_0 > 0$*, then*

$$e^{\ln s} = s.$$

SOLUTION. (a)

$$\begin{aligned}\frac{d}{d\alpha}\ln s(\alpha) &= \frac{s_0'(\alpha)}{s_0(\alpha)} + \sum_{n=1}^{D}(-1)^{n-1}\frac{s_1(\alpha)^{n-1}}{s_0(\alpha)^n}s_1'(\alpha) - \sum_{n=1}^{D}(-1)^{n-1}\frac{s_1(\alpha)^n}{s_0(\alpha)^{n+1}}s_0'(\alpha)\\ &= \frac{s_0'(\alpha)}{s_0(\alpha)} + \sum_{n=0}^{D-1}(-1)^n\frac{s_1(\alpha)^n}{s_0(\alpha)^{n+1}}s_1'(\alpha) + \sum_{n=1}^{D}(-1)^n\frac{s_1(\alpha)^n}{s_0(\alpha)^{n+1}}s_0'(\alpha)\\ &= \frac{s_0'(\alpha)}{s_0(\alpha)} + \frac{s_1'(\alpha)}{s_0(\alpha)} + \sum_{n=1}^{D}(-1)^n\frac{s_1(\alpha)^n}{s_0(\alpha)^{n+1}}[s_1'(\alpha)+s_0'(\alpha)]\end{aligned}$$

since $s_1(\alpha)^D s_1'(\alpha) = \big(1/(D+1)\big)ds_1(\alpha)^{D+1}/d\alpha = 0$. By Problem 1.1,

$$\frac{d}{d\alpha}\ln s(\alpha) = \frac{s_0'(\alpha)}{s_0(\alpha)} + \frac{s_1'(\alpha)}{s_0(\alpha)} + \left[\frac{1}{s(\alpha)} - \frac{1}{s_0(\alpha)}\right][s_1'(\alpha)+s_0'(\alpha)] = \frac{s'(\alpha)}{s(\alpha)}.$$

(b) For the first part use (with $\alpha \in \mathbb{R}$)

$$\frac{d}{d\alpha}\ln e^{\alpha s} = \frac{se^{\alpha s}}{e^{\alpha s}} = s = \frac{d}{d\alpha}\alpha s.$$

This implies that $\ln e^{\alpha s} - \alpha s$ is independent of α. As it is zero for $\alpha = 0$, it is zero for all α. For the second part, set $s' = e^{\ln s}$. Then, by the first part, $\ln s' = \ln s$. So it suffices to prove that ln is injective. This is done by expanding out $s' = \sum_{\ell=0}^{D} s_\ell'$ and $s = \sum_{\ell=0}^{D} s_\ell$ with s_ℓ', $s_\ell \in \bigwedge^\ell \mathcal{V}$ and verifying that every $s_\ell' = s_\ell$ by induction on ℓ. Projecting

$$\ln s_0 + \sum_{n=1}^{D}\frac{(-1)^{n-1}}{n}\left(\frac{s-s_0}{s_0}\right)^n = \ln s_0' + \sum_{n=1}^{D}\frac{(-1)^{n-1}}{n}\left(\frac{s'-s_0'}{s_0'}\right)^n$$

onto $\bigwedge^0 \mathcal{V}$ gives $\ln s_0 = \ln s_0'$ and hence $s_0' = s_0$. Projecting both sides onto $\bigwedge^1 \mathcal{V}$ gives

$$\frac{(-1)^{1-1}}{1}\left(\frac{s-s_0}{s_0}\right)_1 = \frac{(-1)^{1-1}}{1}\left(\frac{s'-s_0'}{s_0'}\right)_1$$

(note that $\big((s-s_0')/s_0'\big)^n$ has no component in $\bigwedge^m \mathcal{V}$ for any $m < n$) and hence $s_1' = s_1$. Projecting both sides onto $\bigwedge^2 \mathcal{V}$ gives

$$\frac{(-1)^{1-1}}{1}\left(\frac{s-s_0}{s_0}\right)_2 + \frac{(-1)^{2-1}}{2}\left[\left(\frac{s-s_0}{s_0}\right)_1\right]^2 = \frac{(-1)^{1-1}}{1}\left(\frac{s'-s_0'}{s_0'}\right)_2 + \frac{(-1)^{2-1}}{2}\left[\left(\frac{s'-s_0'}{s_0'}\right)_1\right]^2.$$

Since $s_0' = s_0$ and $s_1' = s_1$,

$$\frac{s^2}{s_0} = \frac{(-1)^{1-1}}{1}\left(\frac{s-s_0}{s_0}\right)_2 = \frac{(-1)^{1-1}}{1}\left(\frac{s'-s_0'}{s_0'}\right)_2 = \frac{s_2'}{s_0'}$$

and consequently $s_2' = s_2$. And so on. □

PROBLEM 1.5. *Use the notation of Problems 1.2 and 1.4. Prove that if $s, t \in \bigwedge \mathcal{V}$ with $st = ts$ and $s_0, t_0 > 0$, then*

$$\ln(st) = \ln s + \ln t.$$

SOLUTION. Let $u = \ln s$ and $v = \ln t$. Then $s = e^u$ and $t = e^v$ and, as $uv = vu$,

$$\ln(st) = \ln(e^u e^v) = \ln(e^{u+v}) = u + v = \ln s + \ln t. \qquad \square$$

PROBLEM 1.6. *Generalize Problems 1.1–1.5 to $\bigwedge_{\mathbb{S}} \mathcal{V}$ with $\mathbb{S}$ a finite-dimensional graded superalgebra having $\mathbb{S}_0 = \mathbb{C}$.*

SOLUTION. The definitions of Problems 1.1–1.5 still make perfectly good sense when the Grassmann algebra $\bigwedge \mathcal{V}$ is replaced by a finite-dimensional graded superalgebra $\mathbb{S}$ having $\mathbb{S}_0 = \mathbb{C}$. Note that, because $\mathbb{S}$ is finite-dimensional, there is a finite $D_{\mathbb{S}}$ such that $\mathbb{S} = \bigoplus_{m=0}^{D_{\mathbb{S}}} \mathbb{S}_m$. Furthermore, as was observed in Definitions 1.5 and 1.6, $\mathbb{S}' = \bigwedge_{\mathbb{S}} \mathcal{V}$ is itself a finite-dimensional graded superalgebra, with $\mathbb{S}' = \oplus_{m=0}^{D_{\mathbb{S}}+D} \mathbb{S}'_m$ where $\mathbb{S}'_m = \bigoplus_{m_1+m_2=m} \mathbb{S}_{m_1} \oplus \bigwedge^{m_2} \mathcal{V}$. In particular $\mathbb{S}'_0 = \mathbb{S}_0$. So, just replace

> Let $\mathcal{V}$ be a complex vector space of dimension D. Every element s of $\bigwedge \mathcal{V}$ has a unique decomposition $s = s_0 + s_1$ with $s_0 \in \mathbb{C}$ and $s_1 \in \bigoplus_{n=1}^{D} \bigwedge^n \mathcal{V}$.

by

> Let $\mathbb{S}' = \bigoplus_{m=0}^{D'} \mathbb{S}'_m$ be a finite-dimensional graded superalgebra having $\mathbb{S}'_0 = \mathbb{C}$. Every element s of $\mathbb{S}'$ has a unique decomposition $s = s_0 + s_1$ with $s_0 \in \mathbb{C}$ and $s_1 \in \bigoplus_{m=1}^{D'} \mathbb{S}'_m$.

The other parts of the statements and even the solutions remain the same, except for trivial changes, like replacing D by D'. $\square$

PROBLEM 1.7. *Let $\mathcal{V}$ be a complex vector space of dimension D. Let $s = s_0 + s_1 \in \bigwedge \mathcal{V}$ with $s_0 \in \mathbb{C}$ and $s_1 \in \bigoplus_{n=1}^{D} \bigwedge^n \mathcal{V}$. Let $f(z)$ be a complex valued function that is analytic in $|z| < r$. Prove that if $|s_0| < r$, then $\sum_{n=0}^{\infty} f^{(n)}(0) s^n / n!$ converges and*

$$\sum_{n=0}^{\infty} \frac{1}{n!} f^{(n)}(0) s^n = \sum_{n=0}^{D} \frac{1}{n!} f^{(n)}(s_0) s_1^n.$$

SOLUTION. The Taylor series $\sum_{m=0}^{\infty} f^{(m)}(0) t^m / m!$ converges absolutely and uniformly to $f(t)$ for all t in any compact subset of $\{z \in \mathbb{C} \mid |z| < r\}$. Hence $\sum_{m=0}^{\infty} f^{(n+m)}(0) s_0^m / m!$ converges to $f^{(n)}(s_0)$ and

$$\begin{aligned}
\sum_{n=0}^{D} \frac{1}{n!} f^{(n)}(s_0) s_1^n &= \sum_{n=0}^{D} \sum_{m=0}^{\infty} \frac{1}{n!m!} f^{(n+m)}(0) s_0^m s_1^n \\
&= \sum_{N=0}^{\infty} \sum_{n=0}^{\min\{N,D\}} \frac{1}{n!(N-n)!} f^{(N)}(0) s_0^{N-n} s_1^n \quad \text{where } N = m + n \\
&= \sum_{N=0}^{\infty} \frac{1}{N!} f^{(N)}(0) \sum_{n=0}^{\min\{N,D\}} \binom{N}{n} s_0^{N-n} s_1^n
\end{aligned}$$

$$= \sum_{N=0}^{\infty} \frac{1}{N!} f^{(N)}(0) \sum_{n=0}^{N} \binom{N}{n} s_0^{N-n} s_1^n$$
$$= \sum_{N=0}^{\infty} \frac{1}{N!} f^{(N)}(0) s^N. \qquad \square$$

PROBLEM 1.8. *Let $a_1, \ldots, a_D$ be an ordered basis for $\mathcal{V}$. Let $b_i = \sum_{j=1}^{D} M_{i,j} a_j$, $1 \le i \le D$ be another ordered basis for $\mathcal{V}$. Prove that*

$$\int \cdot \, da_D \cdots da_1 = \det M \int \cdot \, db_D \cdots db_1 .$$

In particular, if $b_i = a_{\sigma(i)}$ for some permutation $\sigma \in S_D$

$$\int \cdot \, da_D \cdots da_1 = \operatorname{sgn} \sigma \int \cdot \, db_D \cdots db_1 .$$

SOLUTION. By linearity, it suffices to verify that

$$\int b_{i_1} \ldots b_{i_\ell} \, da_D \cdots da_1 = 0$$

unless $\ell = D$ and

$$\int b_1 \cdots b_D \, da_D \cdots da_1 = \det M .$$

That $\int b_{i_1} \cdots b_{i_\ell} \, da_D \cdots da_1 = 0$ when $\ell \ne D$ is obvious, because $\int a_{j_1} \cdots a_{j\ell} \, da_D \times \cdots da_1$ vanishes when $\ell \ne D$.

$$\int b_1 \cdots b_D \, da_D \cdots da_1 = \sum_{j_1, \ldots, j_D = 1}^{D} M_{1,j_1} \cdots M_{D,j_D} \int a_{j_1} \cdots a_{j_D} \, da_D \cdots da_1 .$$

Now $\int a_{j_1} \cdots a_{j_D} \, da_D \cdots da_1 = 0$ unless all of the j_k's are different so

$$\begin{aligned} \int b_1 \cdots b_D \, da_D \cdots da_1 &= \sum_{\pi \in S_D} M_{1,\pi(1)} \cdots M_{D,\pi(D)} \int a_{\pi(1)} \cdots a_{\pi(D)} \, da_D \cdots da_1 \\ &= \sum_{\pi \in S_D} M_{1,\pi(1)} \cdots M_{D,\pi(D)} \int \operatorname{sgn} \pi a_1 \cdots a_D \, da_D \cdots da_1 \\ &= \sum_{\pi \in S_D} \operatorname{sgn} \pi M_{1,\pi(1)} \cdots M_{D,\pi(D)} = \det M . \end{aligned}$$

In particular, if $b_i = a_{\sigma(i)}$ for some permutation $\sigma \in S_D$

$$\begin{aligned} \int b_1 \cdots b_D \, da_D \cdots da_1 &= \int a_{\sigma(1)} \cdots a_{\sigma(D)} \, da_D \cdots da_1 \\ &= \int \operatorname{sgn} \sigma a_1 \cdots a_D \, da_D \cdots da_1 = \operatorname{sgn} \sigma . \qquad \square \end{aligned}$$

PROBLEM 1.9. *Let*

- *S be a matrix*
- *λ be a real number*
- *$\vec{v}_1$ and $\vec{v}_2$ be two mutually perpendicular, complex conjugate unit vectors*
- *$S\vec{v}_1 = \imath\lambda\vec{v}_1$ and $S\vec{v}_2 = -\imath\lambda\vec{v}_2$.*

Set

$$\vec{w}_1 = \frac{1}{\sqrt{2}\imath}(\vec{v}_1 - \vec{v}_2) \quad \vec{w}_2 = \frac{1}{\sqrt{2}}(\vec{v}_1 + \vec{v}_2).$$

(a) *Prove that*

- $\vec{w}_1$ *and* $\vec{w}_2$ *are two mutually perpendicular, real unit vectors*
- $S\vec{w}_1 = \lambda\vec{w}_2$ *and* $S\vec{w}_2 = -\lambda\vec{w}_1$.

(b) *Suppose, in addition, that* S *is a* 2×2 *matrix. Let* R *be the* 2×2 *matrix whose first column is* $\vec{w}_1$ *and whose second column is* $\vec{w}_2$. *Prove that* R *is a real orthogonal matrix and that* $R^tSR = \left[\begin{smallmatrix} 0 & -\lambda \\ \lambda & 0\end{smallmatrix}\right]$

(c) *Generalize to the case in which* S *is a* $2r\times 2r$ *matrix.*

SOLUTION. (a) Aside from a factor of $\sqrt{2}$, $\vec{w}_1$ and $\vec{w}_2$ are the imaginary and real parts of $\vec{v}_1$, so they are real. I will use the convention that the complex conjugate is on the left argument of the dot product. Then

$$\begin{aligned}
\vec{w}_1\cdot\vec{w}_1 &= \tfrac{1}{2}(\vec{v}_1 - \vec{v}_2)\cdot(\vec{v}_1 - \vec{v}_2) = \tfrac{1}{2}(\vec{v}_1\cdot\vec{v}_1 + \vec{v}_2\cdot\vec{v}_2) = 1\\
\vec{w}_2\cdot\vec{w}_2 &= \tfrac{1}{2}(\vec{v}_1 + \vec{v}_2)\cdot(\vec{v}_1 + \vec{v}_2) = \tfrac{1}{2}(\vec{v}_1\cdot\vec{v}_1 + \vec{v}_2\cdot\vec{v}_2) = 1\\
\vec{w}_1\cdot\vec{w}_2 &= \tfrac{\imath}{2}(\vec{v}_1 - \vec{v}_2)\cdot(\vec{v}_1 + \vec{v}_2) = \tfrac{\imath}{2}(\vec{v}_1\cdot\vec{v}_1 - \vec{v}_2\cdot\vec{v}_2) = 0
\end{aligned}$$

and

$$\begin{aligned}
S\vec{w}_1 &= \frac{1}{\sqrt{2}\imath}(S\vec{v}_1 - S\vec{v}_2) = \frac{1}{\sqrt{2}\imath}(\imath\lambda\vec{v}_1 + \imath\lambda\vec{v}_2) = \frac{\lambda}{\sqrt{2}}(\vec{v}_1 + \vec{v}_2) = \lambda\vec{w}_2\\
S\vec{w}_2 &= \frac{1}{\sqrt{2}}(S\vec{v}_1 + S\vec{v}_2) = \frac{1}{\sqrt{2}}(\imath\lambda\vec{v}_1 - \imath\lambda\vec{v}_2) = \frac{\imath\lambda}{\sqrt{2}}(\vec{v}_1 - \vec{v}_2) = -\lambda\vec{w}_1.
\end{aligned}$$

(b) The matrix R is real, because its column vectors $\vec{w}_1$ and $\vec{w}_2$ are real and it is orthogonal because its column vectors are mutually perpendicular unit vectors. Furthermore,

$$\begin{aligned}
\begin{bmatrix}\vec{w}_1^t\\ \vec{w}_2^t\end{bmatrix} S\begin{bmatrix}\vec{w}_1 & \vec{w}_2\end{bmatrix} &= \begin{bmatrix}\vec{w}_1^t\\ \vec{w}_2^t\end{bmatrix}\begin{bmatrix}S\vec{w}_1 & S\vec{w}_2\end{bmatrix} = \begin{bmatrix}\vec{w}_1^t\\ \vec{w}_2^t\end{bmatrix}\begin{bmatrix}\lambda\vec{w}_2 & -\lambda\vec{w}_1\end{bmatrix}\\
&= \lambda\begin{bmatrix}\vec{w}_1\cdot\vec{w}_2 & -\vec{w}_1\cdot\vec{w}_1\\ \vec{w}_2\cdot\vec{w}_2 & -\vec{w}_2\cdot\vec{w}_1\end{bmatrix} = \begin{bmatrix}0 & -\lambda\\ \lambda & 0\end{bmatrix}.
\end{aligned}$$

(c) Let $r\in\mathbb{N}$ and

- S be a $2r\times 2r$ matrix
- λ_ℓ, $1\le\ell\le r$ be real numbers
- for each $1\le\ell\le r$, $\vec{v}_{1,\ell}$ and $\vec{v}_{2,\ell}$ be two mutually perpendicular, complex conjugate unit vectors
- $\vec{v}_{i,\ell}$ and $\vec{v}_{j,\ell'}$ be perpendicular for all $1\le i,j\le 2$ and all $\ell\ne\ell'$ between 1 and r
- $S\vec{v}_{1,\ell} = \imath\lambda_\ell\vec{v}_{1,\ell}$ and $S\vec{v}_{2,\ell} = -\imath\lambda_\ell\vec{v}_{2,\ell}$, for each $1\le\ell\le r$.

Set

$$\vec{w}_{1,\ell} = \frac{1}{\sqrt{2}\imath}(\vec{v}_{1,\ell} - \vec{v}_{2,\ell}) \quad \vec{w}_{2,\ell} = \frac{1}{\sqrt{2}}(\vec{v}_{1,\ell} - \vec{v}_{2,\ell})$$

for each $1\le\ell\le r$. Then, as in part (a), $w_{1,1}$, $w_{2,1}$, $w_{1,2}$, $w_{2,2}$, $\dots$, $w_{1,r}$, $w_{2,r}$ are mutually perpendicular real unit vectors obeying

$$S\vec{w}_{1,\ell} = \lambda_\ell\vec{w}_{2,\ell} \quad S\vec{w}_{2,\ell} = \lambda_\ell\vec{w}_{1,\ell}$$

and if $R = [w_{1,1}, w_{2,1}, w_{1,2}, w_{2,2}, \ldots, w_{1,r}, w_{2,r}]$, then

$$R^t SR = \bigoplus_{\ell=1}^{r} \begin{bmatrix} 0 & -\lambda_\ell \\ \lambda_\ell & 0 \end{bmatrix}. \qquad \square$$

PROBLEM 1.10. *Let $P(z) = \sum_{i\geq 0} c_i z^i$ be a power series with complex coefficients and infinite radius of convergence and let $f(a)$ be an* even *element of $\bigwedge_{\mathbb{S}} \mathcal{V}$. Show that*

$$\frac{\partial}{\partial a_\ell} P\big(f(a)\big) = P'\big(f(a)\big)\Big(\frac{\partial}{\partial a_\ell} f(a)\Big).$$

SOLUTION. Write $f(a) = f_0 + f_1(a)$ with $f_0 \in \mathbb{C}$ and $f_1(a) \in (\mathbb{S}\setminus\mathbb{S}_0)\oplus\bigoplus_{m=1}^{D} \mathbb{S}\otimes \bigwedge^m \mathcal{V}$. We allow $P(z)$ to have any radius of convergence strictly larger than $|f_0|$. By Problem 1.7,

$$\sum_{n=0}^{\infty} c_i f(a)^n = \sum_{n=0}^{D} \frac{1}{n!} P^{(n)}(f_0) f_1(a)^n.$$

As $\partial f(a)/\partial a_\ell = \partial f_1(a)/\partial a_\ell$, it suffices, by linearity in P, to show that

$$\frac{\partial}{\partial a_\ell}\big(f_1(a)\big)^n = n\big(f_1(a)\big)^{n-1}\Big(\frac{\partial}{\partial a_\ell} f_1(a)\Big). \tag{D.1}$$

For any even $g(a)$ and any $h(a)$, the product rule

$$\frac{\partial}{\partial a_\ell}[g(a)h(a)] = \Big[\frac{\partial}{\partial a_\ell} g(a)\Big] h(a) + g(a)\Big[\frac{\partial}{\partial a_\ell} h(a)\Big]$$

applies. (D.1) follows easily from the product rule, by induction on n. $\square$

PROBLEM 1.11. *Let $\mathcal{V}$ and $\mathcal{V}'$ be vector spaces with bases $\{a_1, \ldots, a_D\}$ and $\{b_1, \ldots, b_{D'}\}$ respectively. Let S and T be $D \times D$ and $D' \times D'$ skew symmetric matrices. Prove that*

$$\int\Big[\int f(a,b)\, d\mu_S(a)\Big] d\mu_T(b) = \int\Big[\int f(a,b)\, d\mu_T(b)\Big] d\mu_S(a).$$

SOLUTION. Let $\{c_1, \ldots, c_D\}$ and $\{d_1, \ldots, d_{D'}\}$ be bases for second copies of $\mathcal{V}$ and $\mathcal{V}'$, respectively. It suffices to consider $f(a,b) = e^{\sum_i c_i a_i + \sum_i d_i b_i}$, because all $f(a,b)$'s can be constructed by taking linear combinations of derivatives of $e^{\sum_i c_i a_i + \sum_i d_i b_i}$ with respect to c_i's and d_i's. For $f(a,b) = e^{\sum_i c_i a_i + \sum_i d_i b_i}$

$$\begin{aligned}
\int\Big[\int f(a,b)\, d\mu_S(a)\Big] d\mu_T(b) &= \int\Big[\int e^{\sum_i c_i a_i + \sum_i d_i b_i}\, d\mu_S(a)\Big] d\mu_T(b) \\
&= \int\Big[e^{\sum_i d_i b_i} \int e^{\sum_i c_i a_i}\, d\mu_S(a)\Big] d\mu_T(b) \\
&= \int \Big[e^{\sum_i d_i b_i} e^{-(\sum_{ij} c_i S_{ij} c_j)/2}\Big]\, d\mu_T(b) \\
&= e^{-(\sum_{ij} c_i S_{ij} c_j)/2} \int e^{\sum_i d_i b_i}\, d\mu_T(b) \\
&= e^{-(\sum_{ij} c_i S_{ij} c_j)/2} e^{-(\sum_{ij} d_i T_{ij} d_j)/2}
\end{aligned}$$

and

$$\begin{aligned}\int\Big[\int f(a,b)\,d\mu_T(b)\Big]\,d\mu_S(a) &= \int\Big[\int e^{\sum_i c_ia_i+\sum_i d_ib_i}\,d\mu_T(b)\Big]\,d\mu_S(a)\\ &= \int\Big[e^{\sum_i c_ia_i}\int e^{\sum_i d_ib_i}\,d\mu_T(b)\Big]\,d\mu_S(a)\\ &= \int\Big[e^{\sum_i c_ia_i}e^{-(\sum_{ij}d_iT_{ij}d_j)/2}\Big]\,d\mu_S(a)\\ &= e^{-(\sum_{ij}d_iT_{ij}d_j)/2}\int e^{\sum_i c_ia_i}\,d\mu_S(a)\\ &= e^{-(\sum_{ij}d_iT_{ij}d_j)/2}e^{-(\sum_{ij}c_iS_{ij}c_j)/2}\end{aligned}$$

are equal. □

PROBLEM 1.12. *Let $\mathcal{V}$ be a D-dimensional vector space with basis $\{a_1,\ldots,a_D\}$ and $\mathcal{V}'$ be a second copy of $\mathcal{V}$ with basis $\{c_1,\ldots,c_D\}$. Let S be a $D\times D$ skew symmetric matrix. Prove that*

$$\int e^{\sum_i c_ia_i}f(a)\,d\mu_S(a) = e^{-(\sum_{ij}c_iS_{ij}c_j)/2}\int f(a-Sc)\,d\mu_S(a).$$

Here $(Sc)_i=\sum_j S_{ij}c_j$.

SOLUTION. It suffices to consider $f(a)=e^{\sum_i b_ia_i}$, because all f's can be constructed by taking linear combinations of derivatives of $e^{\sum_i b_ia_i}$ with respect to b_i's. For $f(a)=e^{\sum_i b_ia_i}$,

$$\begin{aligned}\int e^{\sum_i c_ia_i}f(a)\,d\mu_S(a) &= \int e^{\sum_i(b_i+c_i)a_i}\,d\mu_S(a)\\ &= e^{-(\sum_{ij}(b_i+c_i)S_{ij}(b_j+c_j))/2}\\ &= e^{-(\sum_{ij}c_iS_{ij}c_j)/2}e^{-\sum_{ij}b_iS_{ij}c_j}e^{-(\sum_{ij}b_iS_{ij}b_j)/2}\end{aligned}$$

and

$$\begin{aligned}e^{-(\sum_{ij}c_iS_{ij}c_j)/2}\int f(a-Sc)\,d\mu_S(a) &= e^{-(\sum_{ij}c_iS_{ij}c_j)/2}\int e^{\sum_i b_i(a_i-\sum_j S_{ij}c_j)}\,d\mu_S(a)\\ &= e^{-(\sum_{ij}c_iS_{ij}c_j)/2}e^{-\sum_{ij}b_iS_{ij}c_j}e^{-(\sum_{ij}b_iS_{ij}b_j)/2}\end{aligned}$$

are equal. □

PROBLEM 1.13. *Let $\mathcal{V}$ be a complex vector space with even dimension $D=2r$ and basis $\{\psi_1,\ldots,\psi_r,\bar\psi_1,\ldots,\bar\psi_r\}$. Here, $\bar\psi_i$ need not be the complex conjugate of ψ_i. Let A be an $r\times r$ matrix and $\int\cdot\,d\mu_A(\psi,\bar\psi)$ the Grassmann Gaussian integral obeying*

$$\int \psi_i\psi_j\,d\mu_A(\psi,\bar\psi)=\int\bar\psi_i\bar\psi_j\,d\mu_A(\psi,\bar\psi)=0,\qquad \int\psi_i\bar\psi_j\,d\mu_A(\psi,\bar\psi)=A_{ij}.$$

(a) *Prove*

$$\begin{aligned}&\int \psi_{i_n}\cdots\psi_{i_1}\bar\psi_{j_1}\cdots\bar\psi_{j_m}\,d\mu_A(\psi,\bar\psi)\\ &\qquad=\sum_{\ell=1}^m(-1)^{\ell+1}A_{i_1j_\ell}\int\psi_{i_n}\cdots\psi_{i_2}\bar\psi_{j_1}\cdots\not{\bar\psi}_{j_\ell}\cdots\bar\psi_{j_m}\,d\mu_A(\psi,\bar\psi).\end{aligned}$$

Here, the $\not\bar\psi_{j_\ell}$ *signifies that the factor* $\bar\psi_{j_\ell}$ *is omitted from the integrand.*

(b) *Prove that, if* $n \neq m$,

$$\int \psi_{i_n} \dots \psi_{i_1} \bar\psi_{j_1} \dots \bar\psi_{j_m} \, d\mu_A(\psi, \bar\psi) = 0.$$

(c) *Prove that*

$$\int \psi_{i_n} \dots \psi_{i_1} \bar\psi_{j_1} \dots \bar\psi_{j_n} \, d\mu_A(\psi, \bar\psi) = \det[A_{i_k j_\ell}]_{1 \le k, \ell \le n}.$$

(d) *Let* $\mathcal{V}'$ *be a second copy of* $\mathcal{V}$ *with basis* $\{\zeta_1, \dots, \zeta_r, \bar\zeta_1, \dots, \bar\zeta_r\}$. *View* $e^{\sum_i (\bar\zeta_i \psi_i + \bar\psi_i \zeta_i)}$ *as an element of* $\bigwedge_{\bigwedge \mathcal{V}'} \mathcal{V}$. *Prove that*

$$\int e^{\sum_i (\bar\zeta_i \psi_i + \bar\psi_i \zeta_i)} \, d\mu_A(\psi, \bar\psi) = e^{\sum_{i,j} \bar\zeta_i A_{ij} \zeta_j}.$$

SOLUTION. Define

$$a_i(\psi, \bar\psi) = \begin{cases} \psi_i, & \text{if } 1 \le i \le r \\ \bar\psi_{i-r}, & \text{if } r+1 \le i \le 2r \end{cases}$$

and

$$S = \begin{bmatrix} 0 & A \\ -A^t & 0 \end{bmatrix}.$$

Then

$$\int f(a) \, d\mu_S(a) = \int f\big(a(\psi, \bar\psi)\big) \, d\mu_A(\psi, \bar\psi).$$

(a) Translating the integration by parts formula

$$\int a_k f(a) \, d\mu_S(a) = \sum_{\ell=1}^{D} S_{k\ell} \int \frac{\partial}{\partial a_\ell} f(a) \, d\mu_S(a)$$

of Proposition 1.17 into the $\psi, \bar\psi$ language gives

$$\int \psi_k f(\psi, \bar\psi) \, d\mu_A(\psi, \bar\psi) = \sum_{\ell=1}^{r} A_{k\ell} \int \frac{\partial}{\partial \bar\psi_\ell} f(\psi, \bar\psi) \, d\mu_A(\psi, \bar\psi).$$

Now, just apply this formula to

$$\begin{aligned} \int \psi_{i_n} \cdots \psi_{i_1} \bar\psi_{j_1} \cdots \bar\psi_{j_m} \, d\mu_A(\psi, \bar\psi) \\ = (-1)^{(m+1)(n-1)} \int \psi_{i_1} \bar\psi_{j_1} \cdots \bar\psi_{j_m} \psi_{i_n} \cdots \psi_{i_2} \, d\mu_A(\psi, \bar\psi) \end{aligned}$$

with $k = i_1$ and $f(\psi,\bar\psi) = \bar\psi_{j_1}\cdots\bar\psi_{j_m}\psi_{i_n}\cdots\psi_{i_2}$. This gives

$$\begin{aligned}
&\int \psi_{i_n}\cdots\psi_{i_1}\bar\psi_{j_1}\cdots\bar\psi_{j_m}\,d\mu_A(\psi,\bar\psi)\\
&\quad= (-1)^{(m+1)(n-1)}\sum_{\ell=1}^m (-1)^{\ell-1}A_{i_1j_\ell}\int \bar\psi_{j_1}\cdots\not{\bar\psi}_{j_\ell}\cdots\bar\psi_{j_m}\psi_{i_n}\cdots\psi_{i_2}\,d\mu_A(\psi,\bar\psi)\\
&\quad= (-1)^{(m+1)(n-1)+(m-1)(n-1)}\\
&\qquad\quad\times\sum_{\ell=1}^m (-1)^{\ell-1}A_{i_1j_\ell}\int \psi_{i_n}\cdots\psi_{i_2}\bar\psi_{j_1}\cdots\not{\bar\psi}_{j_\ell}\cdots\bar\psi_{j_m}\,d\mu_A(\psi,\bar\psi)\\
&\quad= \sum_{\ell=1}^m (-1)^{\ell-1}A_{i_1j_\ell}\int \psi_{i_n}\cdots\psi_{i_2}\bar\psi_{j_1}\cdots\not{\bar\psi}_{j_\ell}\cdots\bar\psi_{j_m}\,d\mu_A(\psi,\bar\psi)
\end{aligned}$$

as desired

(d) Define

$$b_i(\psi,\bar\psi) = \begin{cases}\bar\zeta_i, & \text{if } 1\le i\le r\\ -\zeta_{i-r}, & \text{if } r+1\le i\le 2r.\end{cases}$$

Then

$$\int e^{\sum_i(\bar\zeta_i\psi_i+\bar\psi_i\zeta_i)}\,d\mu_A(\psi,\bar\psi) = \int e^{\sum_i b_ia_i}\,d\mu_S(a) = e^{-1/2\sum_{i,j}b_iS_{ij}b_j} = e^{\sum_{i,j}\bar\zeta_iA_{ij}\zeta_j}$$

since

$$\frac12\begin{bmatrix}\bar\zeta & -\zeta\end{bmatrix}\begin{bmatrix}0 & A\\ -A^t & 0\end{bmatrix}\begin{bmatrix}\bar\zeta\\ -\zeta\end{bmatrix} = \frac12\begin{bmatrix}\bar\zeta & -\zeta\end{bmatrix}\begin{bmatrix}-A\zeta\\ -A^t\bar\zeta\end{bmatrix} = \frac12[-\bar\zeta A\zeta + \zeta A^t\bar\zeta] = -\bar\zeta A\zeta.$$

(b) Apply $\prod_{\ell=1}^n \partial/\partial\bar\zeta_{i_\ell}$ and $\prod_{\ell=1}^m \partial/\partial\zeta_{j_\ell}$ to the conclusion of (d) and set $\zeta = \bar\zeta = 0$. The resulting left-hand side is, up to a sign, $\int \psi_{i_n}\cdots\psi_{i_1}\bar\psi_{j_1}\cdots\bar\psi_{j_m}d\mu_A(\psi,\bar\psi)$. Unless $m=n$, the right-hand side is zero.

(c) The proof is by induction on n. For $n=1$, by the definition of $d\mu_A$,

$$\int \psi_{i_1}\bar\psi_{j_1}\,d\mu_A(\psi,\bar\psi) = A_{i_1j_1}$$

as desired. If the result is known for $n-1$, then, by part (a),

$$\begin{aligned}
&\int \psi_{i_n}\cdots\psi_{i_1}\bar\psi_{j_1}\cdots\bar\psi_{j_m}\,d\mu_A(\psi,\bar\psi)\\
&\qquad= \sum_{\ell=1}^m (-1)^{\ell+1}A_{i_1j_\ell}\int \psi_{i_n}\ldots\psi_{i_2}\bar\psi_{j_1}\cdots\not{\bar\psi}_{j_\ell}\cdots\bar\psi_{j_m}\,d\mu_A(\psi,\bar\psi)\\
&\qquad= \sum_{\ell=1}^m (-1)^{\ell+1}A_{i_1j_\ell}\det M^{(1,\ell)}
\end{aligned}$$

where $M^{(1,\ell)}$ is the matrix $[A_{i_\alpha j_\beta}]_{1\le\alpha,\beta\le n}$ with row 1 and column ℓ deleted. So expansion along the first row,

$$\det[A_{i_\alpha j_\beta}]_{1\le\alpha,\beta\le n} = \sum_{\ell=1}^m (-1)^{\ell+1}A_{i_1j_\ell}\det M^{(1,\ell)}$$

gives the desired result. □

PROBLEM 1.14. *Prove that*

$$\mathcal{C}(c) = -\frac{1}{2}\sum_{ij} c_i S_{ij} c_j + \mathcal{G}(-Sc)$$

where $(Sc)_i = \sum_j S_{ij} c_j$.

SOLUTION. By Problem 1.12, with $f(a) = e^{W(a)}$, and Problem 1.5,

$$\begin{aligned}
\mathcal{C}(c) &= \log \frac{1}{Z}\int e^{\sum_i c_i a_i} e^{W(a)}\, d\mu_S(a) \\
&= \log\left[\frac{1}{Z} e^{-1/2\sum_{ij} c_i S_{ij} c_j} \int e^{W(a-Sc)}\, d\mu_S(a)\right] \\
&= -\frac{1}{2}\sum_{ij} c_i S_{ij} c_j + \log\left[\frac{1}{Z}\int e^{W(a-Sc)}\, d\mu_S(a)\right] \\
&= -\frac{1}{2}\sum_{ij} c_i S_{ij} c_j + \mathcal{G}(-Sc).
\end{aligned}$$

□

PROBLEM 1.15. *We have normalized* $\mathcal{G}_{J+1}$ *so that* $\mathcal{G}_{J+1}(0) = 0$. *So the ratio* Z_J/Z_{J+1} *in*

$$\mathcal{G}_{J+1}(c) = \log \frac{Z_J}{Z_{J+1}} \int e^{\mathcal{G}_J(c+a)}\, d\mu_{S^{(J+1)}}(a)$$

had better obey

$$\frac{Z_{J+1}}{Z_J} = \int e^{\mathcal{G}_J(a)}\, d\mu_{S^{(J+1)}}(a).$$

Verify by direct computation that this is the case.

SOLUTION. By Proposition 1.21

$$\begin{aligned}
Z_{J+1} &= \int e^{W(a)}\, d\mu_{S^{(\le J+1)}}(a) \\
&= \int\left[\int e^{W(a+c)}\, d\mu_{S^{(\le J)}}(a)\right] d\mu_{S^{(J+1)}}(c) \\
&= \int \left[Z_j e^{\mathcal{G}_J(a)}\right] d\mu_{S^{(J+1)}}(c) \\
&= Z_j \int e^{\mathcal{G}_J(a)}\, d\mu_{S^{(J+1)}}(c).
\end{aligned}$$

□

PROBLEM 1.16. *Prove that*

$$\begin{aligned}
\mathcal{G}_J &= \Omega_{S^{(J)}} \circ \Omega_{S^{(J-1)}} \circ \cdots \circ \Omega_{S^{(1)}}(W) \\
&= \Omega_{S^{(1)}} \circ \Omega_{S^{(2)}} \circ \cdots \circ \Omega_{S^{(J)}}(W).
\end{aligned}$$

SOLUTION. By the definitions of $\mathcal{G}_J$ and $\Omega_S(W)$,

$$\mathcal{G}_J = \Omega_{S^{(\le J)}}(W).$$

Now apply Theorem 1.29. □

PROBLEM 1.17. *Prove that*

$$\begin{aligned}
{:}f(a){:} &= \int f(a+b)\, d\mu_{-S}(b) \\
f(a) &= \int {:}f{:}(a+b)\, d\mu_S(b).
\end{aligned}$$

SOLUTION. It suffices to consider $f(a) = e^{\sum_i c_i a_i}$. For this f

$$\begin{aligned}\int f(a+b)\,d\mu_{-S}(b) &= \int e^{\sum_i c_i(a_i+b_i)}\,d\mu_{-S}(b)\\ &= e^{\sum_i c_i a_i}\int e^{\sum_i c_i b_i}\,d\mu_{-S}(b)\\ &= e^{c_i a_i}e^{(\sum_{i,j} c_i S_{ij} c_j)/2} = {:}f(a){:}\end{aligned}$$

and

$$\begin{aligned}\int {:}f{:}(a+b)\,d\mu_S(b) &= \int e^{c_i(a_i+b_i)}e^{(\sum_{i,j} c_i S_{ij} c_j)/2}\,d\mu_S(b)\\ &= e^{\sum_i c_i a_i}e^{(\sum_{i,j} c_i S_{ij} c_j)/2}\int e^{\sum_i c_i b_i}\,d\mu_S(b)\\ &= e^{\sum_i c_i a_i}e^{(\sum_{i,j} c_i S_{ij} c_j)/2}e^{-(\sum_{i,j} c_i S_{ij} c_j)/2}\\ &= e^{\sum_i c_i a_i} = f(a).\end{aligned}$$

□

PROBLEM 1.18. *Prove that*

$$\frac{\partial}{\partial a_\ell}{:}f(a){:} = {:}\frac{\partial}{\partial a_\ell}f(a){:}.$$

SOLUTION. By linearity, it suffices to consider $f(a) = a_{\rm I}$ with $\rm I = (i_1 \ldots, i_n)$.

$$\begin{aligned}\frac{\partial}{\partial a_\ell}{:}a_{\rm I}{:} &= \frac{\partial}{\partial a_\ell}\frac{\partial}{\partial b_{i_1}}\cdots\frac{\partial}{\partial b_{i_n}}e^{(\sum_{i,j} b_i S_{ij} b_j)/2}e^{\sum_i b_i a_i}\Big|_{b=0}\\ &= (-1)^n\frac{\partial}{\partial b_{i_1}}\cdots\frac{\partial}{\partial b_{i_n}}\frac{\partial}{\partial a_\ell}e^{(\sum_{i,j} b_i S_{ij} b_j)/2}e^{\sum_i b_i a_i}\Big|_{b=0}\\ &= -(-1)^n\frac{\partial}{\partial b_{i_1}}\cdots\frac{\partial}{\partial b_{i_n}}e^{(\sum_{i,j} b_i S_{ij} b_j)/2}b_\ell e^{\sum_i b_i a_i}\Big|_{b=0}.\end{aligned}$$

If $\ell \notin \rm I$, we get zero. Otherwise, by antisymmetry, we may assume that $\ell = i_n$. Then

$$\begin{aligned}\frac{\partial}{\partial a_\ell}{:}a_{\rm I}{:} &= -(-1)^n\frac{\partial}{\partial b_{i_1}}\cdots\frac{\partial}{\partial b_{i_{n-1}}}e^{1/2\sum_{i,j} b_i S_{ij} b_j}e^{\sum_i b_i a_i}\Big|_{b=0}\\ &= (-1)^{n-1}{:}a_{i_1}\cdots a_{i_{n-1}}{:} = {:}\frac{\partial}{\partial a_\ell}a_{\rm I}{:}.\end{aligned}$$

□

PROBLEM 1.19. *Prove that*

$$\int {:}g(a)a_i{:}f(a)\,d\mu_S(a) = \sum_{\ell=1}^{D} S_{i\ell}\int {:}g(a){:}\frac{\partial}{\partial a_\ell}f(a)\,d\mu_S(a).$$

SOLUTION. It suffices to consider $g(a) = a_{\rm I}$ and $f(a) = a_{\rm J}$. Observe that

$$\begin{aligned}&\int e^{\sum_m b_m a_m}e^{(\sum_{m,j} b_m S_{mj} b_j)/2}e^{\sum_m c_m a_m}\,d\mu_S(a)\\ &\qquad= e^{(\sum_{m,j} b_m S_{mj} b_j)/2}e^{-(\sum_{m,j}(b_m+c_m)S_{mj}(b_j+c_j))/2}\\ &\qquad= e^{-\sum_{m,j} b_m S_{mj} c_j}e^{-(\sum_{m,j} c_m S_{mj} c_j)/2}.\end{aligned}$$

Now apply $\partial/\partial b_i$ to both sides

$$\frac{\partial}{\partial b_i}\int e^{\sum_m b_m a_m} e^{(\sum_{m,j} b_m S_{mj} b_j)/2} e^{\sum_m c_m a_m}\, d\mu_S(a)$$
$$\begin{aligned}
&= -\sum_{\ell=1}^{D} S_{i,\ell} c_\ell e^{-\sum b_m S_{mj} c_j} e^{-1/2\sum c_m S_{mj} c_j} \\
&= -\sum_{\ell=1}^{D} S_{i,\ell} c_\ell \int e^{\sum_m b_m a_m} e^{1/2\sum_{m,j} b_m S_{mj} b_j} e^{\sum_m c_m a_m}\, d\mu_S(a) \\
&= \sum_{\ell=1}^{D} S_{i,\ell} \int e^{\sum_m b_m a_m} e^{(\sum_{m,j} b_m S_{mj} b_j)/2} \frac{\partial}{\partial a_\ell} e^{\sum_m c_m a_m}\, d\mu_S(a).
\end{aligned}$$

Applying

$$(-1)^{|\mathrm{J}|}\prod_{j\in \mathrm{I}}\frac{\partial}{\partial b_j}\prod_{k\in \mathrm{J}}\frac{\partial}{\partial c_k}$$

to both sides and setting $b = c = 0$ gives the desired result. The $(-1)^{|\mathrm{J}|}$ is to move $\prod_{k\in \mathrm{J}} \partial/\partial c_k$ past the $\partial/\partial b_i$ on the left and past the $\partial/\partial a_\ell$ on the right. □

PROBLEM 1.20. *Prove that*

$$:f:(a+b) = :f(a+b):_a = :f(a+b):_b.$$

Here $:\cdot:_a$ *means Wick ordering of the* a_i*'s and* $:\cdot:_b$ *means Wick ordering of the* b_i*'s. Precisely, if* $\{a_i\}$, $\{b_i\}$, $\{A_i\}$, $\{B_i\}$ *are bases of four vector spaces, all of the same dimension,*

$$:e^{\sum_i A_i a_i + \sum_i B_i b_i}:_a = e^{\sum_i B_i b_i} :e^{\sum_i A_i a_i}:_a = e^{\sum_i A_i a_i + \sum_i B_i b_i} e^{(\sum_{ij} A_i S_{ij} A_j)/2}$$
$$:e^{\sum_i A_i a_i + \sum_i B_i b_i}:_a = e^{\sum_i A_i a_i} :e^{\sum_i B_i b_i}_a = e^{\sum_i A_i a_i + \sum_i B_i b_i} e^{(\sum_{ij} B_i S_{ij} B_j)/2}.$$

SOLUTION. It suffices to consider $f(a) = e^{\sum_i c_i a_i}$. Then

$$\begin{aligned}
:f:(a+b) = :f(d):_d\big|_{d=a+b} &= e^{\sum_i c_i d_i} e^{(\sum_{ij} c_i S_{ij} c_j)/2}\big|_{d=a+b} \\
&= e^{\sum_i c_i (a_i+b_i)} e^{(\sum_{ij} c_i S_{ij} c_j)/2}
\end{aligned}$$

and

$$\begin{aligned}
:f(a+b):_a &= :e^{\sum_i c_i(a_i+b_i)}:_a = e^{\sum_i c_i b_i} :e^{\sum_i c_i a_i}:_a \\
&= e^{\sum_i c_i b_i} e^{\sum_i c_i a_i} e^{(\sum_{ij} c_i S_{ij} c_j)/2} = e^{\sum_i c_i(a_i+b_i)} e^{(\sum_{ij} c_i S_{ij} c_j)/2}
\end{aligned}$$

and

$$\begin{aligned}
:f(a+b):_b &= :e^{\sum_i c_i(a_i+b_i)}:_b = e^{\sum_i c_i a_i} :e^{\sum_i c_i b_i}:_b \\
&= e^{\sum_i c_i a_i} e^{\sum_i c_i b_i} e^{(\sum_{ij} c_i S_{ij} c_j)/2} = e^{\sum_i c_i(a_i+b_i)} e^{(\sum_{ij} c_i S_{ij} c_j)/2}. \quad \square
\end{aligned}$$

All three are the same.

PROBLEM 1.21. *Prove that*

$$:f:_{S+T}(a+b) = :f(a+b):_{\substack{a,S\\ b,T}}.$$

Here S *and* T *are skew symmetric matrices,* $:\cdot:_{S+T}$ *means Wick ordering with respect to* $S+T$ *and* $:\cdot:_{\substack{a,S\\ b,T}}$ *means Wick ordering of the* a_i*'s with respect to* S *and of the* b_i*'s with respect to* T.

SOLUTION. It suffices to consider $f(a) = e^{\sum_i c_i a_i}$. Then

$$\begin{aligned} {:}f{:}_{S+T}(a+b) &= {:}f(d){:}_{d,S+T}\big|_{d=a+b} = e^{\sum_i c_i d_i} e^{(\sum_{ij} c_i(S_{ij}+T_{ij})c_j)/2}\big|_{d=a+b} \\ &= e^{\sum_i c_i(a_i+b_i)} e^{(\sum_{ij} c_i(S_{ij}+T_{ij}))/2} c_j \end{aligned}$$

and

$$\begin{aligned} {:}f(a+b){:}_{\substack{a,S\\ b,T}} &= {:}e^{\sum_i c_i(a_i+b_i)}{:}_{\substack{a,S\\ b,T}} = {:}e^{\sum_i c_i a_i}{:}_{a,S}{:}e^{\sum_i c_i b_i}{:}_{b,T} \\ &= e^{\sum_i c_i a_i} e^{(\sum_{ij} c_i S_{ij} c_j)/2} e^{\sum_i c_i b_i} e^{(\sum_{ij} c_i T_{ij} c_j)/2} \\ &= e^{\sum_i c_i(a_i+b_i)} e^{(\sum_{ij} c_i(S_{ij}+T_{ij})c_j)/2}. \end{aligned}$$

The two right-hand sides are the same. □

PROBLEM 1.22. *Prove that*

$$\int {:}f(a){:}\, d\mu_S(a) = f(0).$$

SOLUTION. It suffices to consider $f(a) = e^{\sum_i c_i a_i}$. Then

$$\begin{aligned} \int {:}f(a){:}\, d\mu_S(a) = \int {:}e^{\sum_i c_i a_i}{:}\, d\mu_S(a) &= \int e^{\sum_i c_i a_i} e^{(\sum_{ij} c_i S_{ij} c_j)/2}\, d\mu_S(a) \\ &= e^{(\sum_{ij} c_i S_{ij} c_j)/2} \int e^{\sum_i c_i a_i}\, d\mu_S(a) \\ &= e^{(\sum_{ij} c_i S_{ij} c_j)/2} e^{-(\sum_{ij} c_i S_{ij} c_j)/2} = 1 = f(0). \end{aligned}$$ □

PROBLEM 1.23. *Prove that*

$$\int \prod_{i=1}^{n} {:}\prod_{\mu=1}^{e_i} a_{\ell_{i,\mu}}{:}\, d\mu_S(\psi) = \mathrm{Pf}\big(T_{(i,\mu),(i',\mu')}\big)$$

where

$$T_{(i,\mu),(i',\mu')} = \begin{cases} 0, & \text{if } i = i' \\ S_{\ell_{i,\mu},\ell_{i',\mu'}} & \text{if } i \neq i'. \end{cases}$$

Here T is a skew symmetric matrix with $\sum_{i=1}^{n} e_i$ rows and columns, numbered, in order $(1,1), \ldots, (1,e_1), (2,1), \ldots, (2,e_2), \ldots, (n,e_n)$. The product in the integrand is also in this order.

SOLUTION. The proof is by induction on the number of a's. That is, on $\sum_{\ell=1}^{n} e_\ell$. As

$$\int {:}a_i{:}\, d\mu_S(\psi) = \int a_i\, d\mu_S(\psi) = 0$$

$$\int {:}a_{i_1}{:}{:}a_{i_2}{:}\, d\mu_S(\psi) = \int a_{i_1} a_{i_2}\, d\mu_S(\psi) = S_{i_1,i_2}$$

$$\int {:}a_{i_1} a_{i_2}{:}\, d\mu_S(\psi) = 0$$

the induction starts OK. By Problem 1.19,

$$\int \prod_{i=1}^{n} {:}\prod_{\mu=1}^{e_i} a_{\ell_{i,\mu}}{:}\, d\mu_S(\psi) = \sum_p S_{\ell_{1e_1},p} \int {:}\prod_{\mu=1}^{e_i-1} a_{\ell_{1,\mu}}{:} \frac{\partial}{\partial a_p} \prod_{i=2}^{n} {:}\prod_{\mu=1}^{e_i} a_{\ell_{i,\mu}}{:}\, d\mu_S(\psi).$$

Define

$$E_0 = 0, \qquad E_i = \sum_{j=1}^{i} e_j \qquad \text{and} \qquad l_{E_{i-1}+\mu} = \ell_{i,\mu}.$$

Under this new indexing, the last equation becomes

$$\begin{aligned}\int \prod_{i=1}^{n} :\prod_{j=E_{i-1}+1}^{E_i} a_{l_j}: d\mu_S(\psi) &= \sum_p S_{l_{E_1},p} \int :\prod_{j=1}^{E_1-1} a_{l_j}: \frac{\partial}{\partial a_p} \prod_{i=2}^{n} :\prod_{j=E_{i-1}+1}^{E_i} a_{l_j}: d\mu_S(\psi) \\ &= \sum_{k=E_1+1}^{E_n} (-1)^{k-E_1-1} S_{l_{E_1},l_k} \int :\prod_{j=1}^{E_i-1} a_{l_j}: \prod_{i=2}^{n} :\prod_{\substack{j=E_{i-1}+1 \\ j\neq k}}^{E_i} a_{l_j}: d\mu_S(\psi)\end{aligned}$$

where we have used the product rule (Proposition 1.13), Problem 1.18 and the observation that $:\prod_{\mu=1}^{e_i} a_{\ell_{i,\mu}}:$ has the same parity, even or odd, as e_i does. Let $M_{k\ell}$ be the matrix obtained from T by deleting rows k and ℓ and columns k and ℓ. By the inductive hypothesis

$$\int \prod_{i=1}^{n} :\prod_{j=E_{i-1}+1}^{E_i} a_{l_j}: d\mu_S(\psi) = \sum_{k=E_1+1}^{E_n} (-1)^{k-E_1-1} S_{l_{E_1},l_k} \operatorname{Pf}(M_{E_1k}).$$

By Proposition 1.18.a, with k by E_1,

$$\int \prod_{i=1}^{n} :\prod_{j=E_{i-1}+1}^{E_i} a_{l_j}: d\mu_S(\psi) = \operatorname{Pf}(T)$$

since, for all $k > E_1$, $\operatorname{sgn}(E_1 - k)(-1)^{E_1+k} = (-1)^{k-E_1-1}$. □

PROBLEM 1.24. *Let a_{ij}, $1 \le i,j \le n$ be complex numbers. Prove Hadamard's inequality*

$$|\det[a_{ij}]| \le \prod_{i=1}^{n} \Big(\sum_{j=1}^{n} |a_{ij}|^2 \Big)^{1/2}$$

from Gram's inequality.

SOLUTION. Set $a_i = (a_{i1}, \dots, a_{in})$ and write $a_{ij} = \langle a_i, e_j \rangle$ with the vectors $e_j = (0, \dots, 1, \dots, 0)$, $j = 1, \dots, n$ forming the standard basis for $\mathbb{C}^n$. Then, by Gram's inequality

$$|\det[a_{ij}]| = |\det[\langle a_i, e_j\rangle]| \le \prod_{i=1}^{n} \|a_i\|\,\|e_j\| = \prod_{i=1}^{n} \Big(\sum_{j=1}^{n} |a_{ij}|^2 \Big)^{1/2}. \qquad \square$$

PROBLEM 1.25. *Let $\mathcal{V}$ be a vector space with basis $\{a_1, \dots, a_D\}$. Let $S_{\ell,\ell'}$ be a skew symmetric $D \times D$ matrix with*

$$S(\ell, \ell') = \langle f_\ell, g_{\ell'} \rangle_{\mathcal{H}} \qquad \text{for all } 1 \le \ell, \ell' \le D$$

for some Hilbert space $\mathcal{H}$ and vectors $f_\ell, g_{\ell'} \in \mathcal{H}$. Set $F_\ell = \sqrt{\|f_\ell\|_{\mathcal{H}} \|g_\ell\|_{\mathcal{H}}}$. Prove that

$$\Big| \int \prod_{\ell=1}^{n} a_{i_\ell} \, d\mu_S(a) \Big| \le \prod_{1 \le \ell \le n} F_{i_\ell}.$$

SOLUTION. I will prove both

(a) $$\left|\int \prod_{\ell=1}^{n} a_{i_\ell}\, d\mu_S(a)\right| \le \prod_{\ell=1}^{n} F_{i_\ell}$$

(b) $$\left|\int \prod_{k=1}^{m} a_{i_k} {:}\prod_{\ell=1}^{n} a_{j_\ell}{:}\, d\mu_S(a)\right| \le 2^n \prod_{k=1}^{m} F_{i_k} \prod_{\ell=1}^{n} F_{j_\ell}.$$

For (a) just apply Gram's inequality to

$$\left|\int \prod_{\ell=1}^{n} a_{i_\ell}\, d\mu_S(a)\right| = \left|\mathrm{Pf}[S_{i_k,i_\ell}]_{1\le k,\ell\le n}\right| = \sqrt{\left|\det[S_{i_k,i_\ell}]\right|}.$$

For (b) observe that, by Problem 1.17

$$\begin{aligned}\int \prod_{k=1}^{m} a_{i_k} {:}\prod_{\ell=1}^{n} a_{j_\ell}{:}\, d\mu_S(a) &= \int \prod_{k=1}^{m} a_{i_k} \prod_{\ell=1}^{n} [a_{j_\ell} + b_{j_\ell}]\, d\mu_S(a)\, d\mu_{-S}(b) \\ &= \sum_{\mathrm{I}\subset\{1,\dots,n\}} \pm \int \prod_{k=1}^{m} a_{i_k} \prod_{\ell\in \mathrm{I}} a_{j_\ell}\, d\mu_S(a) \int \prod_{\ell\notin \mathrm{I}} b_{j_\ell}\, d\mu_{-S}(b).\end{aligned}$$

There are 2^n terms. Apply (a) to both factors of each term. □

PROBLEM 1.26. *Prove, under the hypotheses of Proposition* 1.34, *that*

$$\left|\int \prod_{i=1}^{n} {:}\prod_{\mu=1}^{e_i} \psi(\ell_{i,\mu},\kappa_{i,\mu}){:}\, d\mu_A(\psi)\right| \le \prod_{\substack{1\le i\le n\\ 1\le\mu\le e_i\\ \kappa_{i,\mu}=0}} \sqrt{2}\|f_{\ell_{i,\mu}}\|_{\mathcal{H}} \prod_{\substack{1\le i\le n\\ 1\le\mu\le e_i\\ \kappa_{i,\mu}=1}} \sqrt{2}\|g_{\ell_{i,\mu}}\|_{\mathcal{H}}.$$

SOLUTION. Define

$$\begin{aligned} S &= \{(i,\mu) \mid 1\le i\le n, 1\le\mu\le e_i, \kappa_{i,\mu}=0\} \\ \bar{S} &= \{(i,\mu) \mid 1\le i\le n, 1\le\mu\le e_i, \kappa_{i,\mu}=1\}.\end{aligned}$$

If the integral does not vanish, the cardinality of S and $\bar{S}$ coincide and there is a sign $\pm$ such that (this is a special case of Problem 1.23)

$$\int \prod_{i=1}^{n} {:}\prod_{\mu=1}^{e_i} \psi(\ell_{i,\mu},\kappa_{i,\mu}){:}\, d\mu_A(\psi) = \pm\det(M_{\alpha,\beta})_{\substack{\alpha\in S\\ \beta\in\bar{S}}}$$

where

$$M_{(i,\mu),(i',\mu')} = \begin{cases} 0 & \text{if } i=i' \\ A(\ell_{i,\mu},\ell_{i',\mu'}) & \text{if } i\ne i' \end{cases}.$$

Define the vectors u^α, $\alpha\in S$ and v^β, $\beta\in\bar{S}$ in $\mathbb{C}^{n+1}$ by

$$u_i^\alpha = \begin{cases} 1, & \text{if } i=n+1 \\ 1, & \text{if } \alpha=(i,\mu) \text{ for some } 1\le\mu\le e_i \\ 0, & \text{otherwise}\end{cases}$$

$$v_i^\beta = \begin{cases} 1, & \text{if } i=n+1 \\ -1, & \text{if } \beta=(i,\mu) \text{ for some } 1\le\mu\le e_i \\ 0, & \text{otherwise}\end{cases}.$$

Observe that, for all $\alpha \in S$ and $\beta \in \bar{S}$,

$$\|u^\alpha\| = \|v^\beta\| = \sqrt{2}$$

$$u^\alpha \cdot v^\beta = \begin{cases} 1, & \text{if } \alpha = (i,\mu),\ \beta = (i',\mu') \text{ with } i \neq i' \\ 0, & \text{if } \alpha = (i,\mu),\ \beta = (i',\mu') \text{ with } i = i' \end{cases}.$$

Hence, setting

$$F_\alpha = u^\alpha \otimes f_{\ell_{i,\mu}} \in \mathbb{C}^{n+1} \otimes \mathcal{H} \quad \text{for } \alpha = (i,\mu) \in S$$
$$G_\beta = v^\beta \otimes g_{\ell i,\mu} \in \mathbb{C}^{n+1} \otimes \mathcal{H} \quad \text{for } \beta = (i,\mu) \in \bar{S}$$

we have

$$M_{\alpha,\beta} = \langle F_\alpha, G_\beta \rangle_{\mathbb{C}^{n+1}\otimes\mathcal{H}}$$

and consequently, by Gram's inequality,

$$\begin{aligned}\left| \int \prod_{i=1}^{n} : \prod_{\mu=1}^{e_i} \psi(\ell_{i,\mu}, \kappa_{i,\mu}) : d\mu_A(\psi) \right| &= \left| \det(M_{\alpha,\beta})_{\substack{\alpha\in S\\ \beta\in\bar{S}}} \right| \\ &\leq \prod_{\alpha\in S} \|F_\alpha\|_{\mathbb{C}^{n+1}\otimes\mathcal{H}} \prod_{\beta\in\bar{S}} \|G_\beta\|_{\mathbb{C}^{n+1}\otimes\mathcal{H}} \\ &\leq \prod_{\alpha\in S} \sqrt{2}\|f_{\ell_\alpha}\|_{\mathcal{H}} \prod_{\beta\in\bar{S}} \sqrt{2}\|g_{\ell_\beta}\|_{\mathcal{H}}. \end{aligned}$$ □

Chapter 2. Fermionic Expansions

PROBLEM 2.1. *Let*

$$F(a) = \sum_{j_1,j_2=1}^{D} f(j_1,j_2) a_{j_1} a_{j_2}$$
$$W(a) = \sum_{j_1,j_2,j_3,j_4=1}^{D} w(j_1,j_2,j_3,j_4) a_{j_1} a_{j_2} a_{j_3} a_{j_4}$$

with $f(j_1,j_2)$ and $w(j_1,j_2,j_3,j_4)$ antisymmetric under permutation of their arguments.

(a) *Set*

$$\mathcal{S}(\lambda) = \frac{1}{\mathcal{Z}_\lambda} \int F(a) e^{\lambda W(a)}\, d\mu_S(a) \quad \textit{where } \mathcal{Z}_\lambda = \int e^{\lambda W(a)}\, d\mu_S(a).$$

Compute $d^\ell \mathcal{S}(\lambda)/d\lambda^\ell|_{\lambda=0}$ for $\ell = 0,1,2$.

(b) *Set*

$$R(\lambda) = \int :e^{\lambda W(a+b)-\lambda W(a)} - 1:_b F(b)\, d\mu_S(b).$$

Compute $d^\ell R(\lambda)/d\lambda^\ell|_{\lambda=0}$ for all $\ell \in \mathbb{N}$.

SOLUTION. (a) We have defined

$$\mathcal{S}(\lambda) = \frac{\int F(a) e^{\lambda W(a)}\, d\mu_S(a)}{\int e^{\lambda W(a)}\, d\mu_S(a)} = \sum_{j_1,j_2=1}^{D} f(j_1,j_2) \frac{\int a_{j_1} a_{j_2} e^{\lambda W(a)}\, d\mu_S(a)}{\int e^{\lambda W(a)}\, d\mu_S(a)}.$$

Applying Proposition 1.17 (integration by parts) to the numerator with $k = j_1$ and using the antisymmetry of w gives

$$\text{(D.2)}\quad \begin{aligned} \mathcal{S}(\lambda) &= \sum_{j_1,j_2=1}^{D} f(j_1,j_2) \sum_k S_{j_1,k} \frac{\int \partial(a_{j_2} e^{\lambda W(a)})/\partial a_k \, d\mu_S(a)}{\int e^{\lambda W(a)} \, d\mu_S(a)} \\ &= \sum_{j_1,j_2=1}^{D} f(j_1,j_2) \left[S_{j_1,j_2} - \sum_k S_{j_1,k} \frac{\int a_{j_2} e^{\lambda W(a)} \partial\big(\lambda W(a)\big)/\partial a_k \, d\mu_S(a)}{\int e^{\lambda W(a)} \, d\mu_S(a)} \right] \\ &= \sum_{j_1,j_2=1}^{D} f(j_1,j_2) \left[S_{j_1,j_2} - 4\lambda \sum_{j_3,j_4,j_5,j_6} S_{j_1,j_3} w(j_3,j_4,j_5,j_6) \right. \\ &\qquad\qquad \left. \times \frac{\int a_{j_2} a_{j_4} a_{j_5} a_{j_6} e^{\lambda W(a)} \, d\mu_S(a)}{\int e^{\lambda W(a)} \, d\mu_S(a)} \right]. \end{aligned}$$

Setting $\lambda = 0$ gives

$$\mathcal{S}(0) = \sum_{j_1,j_2=1}^{D} f(j_1,j_2) S_{j_1,j_2}.$$

Differentiating once with respect to λ and setting $\lambda = 0$ gives

$$\mathcal{S}'(0) = -4 \sum_{j_1,\cdots,j_6} f(j_1,j_2) S_{j_1,j_3} w(j_3,j_4,j_5,j_6) \int a_{j_2} a_{j_4} a_{j_5} a_{j_6} \, d\mu_S(a).$$

Integrating by parts and using the antisymmetry of w a second time

$$\begin{aligned} \mathcal{S}'(0) &= -12 \sum_{j_1,\cdots,j_6} f(j_1,j_2) S_{j_1,j_3} S_{j_2,j_4} w(j_3,j_4,j_5,j_6) \int a_{j_5} a_{j_6} \, d\mu_S(a) \\ &= -12 \sum_{j_1,\cdots,j_6} f(j_1,j_2) S_{j_1,j_3} S_{j_2,j_4} w(j_3,j_4,j_5,j_6) S_{j_5,j_6}. \end{aligned}$$

To determine $\mathcal{S}''(0)$ we need the order λ contribution to

$$\begin{aligned} &\sum_{j_3,j_4,j_5,j_6} S_{j_1,j_3} w(j_3,j_4,j_5,j_6) \frac{\int a_{j_2} a_{j_4} a_{j_5} a_{j_6} e^{\lambda W(a)} \, d\mu_S(a)}{\int e^{\lambda W(a)} \, d\mu_S(a)} \\ &= 3 \sum_{j_3,j_4,j_5,j_6} S_{j_1,j_3} S_{j_2,j_4} w(j_3,j_4,j_5,j_6) \frac{\int a_{j_5} a_{j_6} e^{\lambda W(a)} \, d\mu_S(a)}{\int e^{\lambda W(a)} \, d\mu_S(a)} \\ &\quad - \sum_{j_3,j_4,j_5,j_6,k} S_{j_1,j_3} w(j_3,j_4,j_5,j_6) S_{j_2,k} \frac{\int a_{j_4} a_{j_5} a_{j_6} e^{\lambda W(a)} \partial\big(\lambda W(a)\big)/\partial a_k \, d\mu_S(a)}{\int e^{\lambda W(a)} \, d\mu_S(a)} \\ &= 3 \sum_{j_3,j_4,j_5,j_6} S_{j_1,j_3} S_{j_2,j_4} w(j_3,j_4,j_5,j_6) S_{j_5,j_6} \\ &\quad - 3 \sum_{j_3,j_4,j_5,j_6,k} S_{j_1,j_3} S_{j_2,j_4} w(j_3,j_4,j_5,j_6) S_{j_5,k} \frac{\int a_{j_6} e^{\lambda W(a)} \partial\big(\lambda W(a)\big)/\partial a_k \, d\mu_S(a)}{\int e^{\lambda W(a)} \, d\mu_S(a)} \\ &\quad - \sum_{j_3,j_4,j_5,j_6,k} S_{j_1,j_3} w(j_3,j_4,j_5,j_6) S_{j_2,k} \frac{\int a_{j_4} a_{j_5} a_{j_6} e^{\lambda W(a)} \partial\big(\lambda W(a)\big)/\partial a_k \, d\mu_S(a)}{\int e^{\lambda W(a)} \, d\mu_S(a)} \\ &= 3 \sum_{j_3,j_4,j_5,j_6} S_{j_1,j_3} S_{j_2,j_4} w(j_3,j_4,j_5,j_6) S_{j_5,j_6} \end{aligned}$$

$$- 3\lambda \sum_{j_3,j_4,j_5,j_6,k} S_{j_1,j_3} S_{j_2,j_4} w(j_3,j_4,j_5,j_6) S_{j_5,k} \int a_{j_6} \frac{\partial}{\partial a_k} W(a)\, d\mu_S(a)$$
$$- \lambda \sum_{j_3,j_4,j_5,j_6,k} S_{j_1,j_3} w(j_3,j_4,j_5,j_6) S_{j_2,k} \int a_{j_4} a_{j_5} a_{j_6} \frac{\partial}{\partial a_k} W(a)\, d\mu_S(a) + \mathcal{O}(\lambda^2)$$

$$\begin{aligned}
&S''(0)\\
&= 24 \sum_{j_1,j_2=1}^{D} f(j_1,j_2) \Big[\sum_{j_3,j_4,j_5,j_6,k} S_{j_1,j_3} S_{j_2,j_4} w(j_3,j_4,j_5,j_6) S_{j_5,k} \int a_{j_6} \frac{\partial}{\partial a_k} W(a)\, d\mu_S(a) \Big]\\
&\quad + 8 \sum_{j_1,j_2=1}^{D} f(j_1,j_2) \Big[\sum_{j_3,j_4,j_5,j_6,k} S_{j_1,j_3} w(j_3,j_4,j_5,j_6) S_{j_2,k} \int a_{j_4} a_{j_5} a_{j_6} \frac{\partial}{\partial a_k} W(a)\, d\mu_S(a) \Big]\\
&= 24 \sum_{j_1,j_2=1}^{D} f(j_1,j_2) \Big[\sum_{j_3,j_4,j_5,j_6,k} S_{j_1,j_3} S_{j_2,j_4} w(j_3,j_4,j_5,j_6) S_{j_5,k} \int a_{j_6} \frac{\partial}{\partial a_k} W(a)\, d\mu_S(a) \Big]\\
&\quad + 16 \sum_{j_1,j_2=1}^{D} f(j_1,j_2) \Big[\sum_{j_3,j_4,j_5,j_6,k} S_{j_1,j_3} w(j_3,j_4,j_5,j_6) S_{j_4,j_5} S_{j_2,k} \int a_{j_6} \frac{\partial}{\partial a_k} W(a)\, d\mu_S(a) \Big]\\
&\quad + 8 \sum_{j_1,j_2=1}^{D} f(j_1,j_2) \sum_{\substack{j_3,\dots,j_6\\ k,\ell}} S_{j_1,j_3} w(j_3,j_4,j_5,j_6) S_{j_2,k} S_{j_4,\ell} \int a_{j_5} a_{j_6} \frac{\partial}{\partial a_\ell} \frac{\partial}{\partial a_k} W(a)\, d\mu_S(a) \Big]\\
&= 24 \cdot 4 \cdot 3 \sum_{j_1,\dots,j_{10}} f(j_1,j_2) S_{j_1,j_3} S_{j_2,j_4} w(j_3,j_4,j_5,j_6) S_{j_5,j_7} S_{j_6,j_8} w(j_7,j_8,j_9,j_{10}) S_{j_9,j_{10}}\\
&\quad + 16 \cdot 4 \cdot 3 \sum_{j_1,\dots,j_{10}} f(j_1,j_2) S_{j_1,j_3} S_{j_2,j_7} w(j_3,j_4,j_5,j_6) S_{j_4,j_5} S_{j_6,j_8} w(j_7,j_8,j_9,j_{10}) S_{j_9,j_{10}}\\
&\quad + 8 \cdot 4 \cdot 3 \sum_{j_1,\dots,j_{10}} f(j_1,j_2) S_{j_1,j_3} S_{j_2,j_7} w(j_3,j_4,j_5,j_6) S_{j_4,j_8} S_{j_5,j_6} w(j_7,j_8,j_0,j_{10}) S_{j_9,j_{10}}\\
&\quad - 8 \cdot 4! \sum_{j_1,\dots,j_{10}} f(j_1,j_2) S_{j_1,j_3} S_{j_2,j_7} w(j_3,j_4,j_5,j_6) S_{j_4,j_8} S_{j_5,j_9} S_{j_6,j_{10}} w(j_7,j_8,j_9,j_{10})\\
&= 24 \cdot 4 \cdot 3 \sum_{j_1,\dots,j_{10}} f(j_1,j_2) S_{j_1,j_3} S_{j_2,j_4} w(j_3,j_4,j_5,j_6) S_{j_5,j_7} S_{j_6,j_8} w(j_7,j_8,j_9,j_{10}) S_{j_9,j_{10}}\\
&\quad + 24 \cdot 4 \cdot 3 \sum_{j_1,\dots,j_{10}} f(j_1,j_2) S_{j_1,j_3} S_{j_2,j_7} w(j_3,j_4,j_5,j_6) S_{j_4,j_5} S_{j_6,j_8} w(j_7,j_8,j_9,j_{10}) S_{j_9,j_{10}}\\
&\quad - 8 \cdot 4! \sum_{j_1,\dots,j_{10}} f(j_1,j_2) S_{j_1,j_3} S_{j_2,j_7} w(j_3,j_4,j_5,j_6) S_{j_4,j_8} S_{j_5,j_9} S_{j_6,j_{10}} w(j_7,j_8,j_9,j_{10}).
\end{aligned}$$

By way of resume, the answer to part (a) is

$$\mathcal{S}(0) = \sum_{j_1,j_2=1}^{D} f(j_1,j_2) S_{j_1,j_2}$$

$$\mathcal{S}'(0) = -12 \sum_{j_1,\dots,j_6=1}^{D} f(j_1,j_2) S_{j_1,j_3} S_{j_2,j_4} w(j_3,j_4,j_5,j_6) S_{j_5,j_6}$$

$$\mathcal{S}''(0) = 288 \sum_{j_1,\cdots,j_{10}} f(j_1,j_2) S_{j_1,j_3} S_{j_2,j_4} w(j_3,j_4,j_5,j_6) S_{j_5,j_7} S_{j_6,j_8} w(j_7,j_8,j_9,j_{10}) S_{j_9,j_{10}}$$
$$+ 288 \sum_{j_1,\cdots,j_{10}} f(j_1,j_2) S_{j_1,j_3} S_{j_2,j_7} w(j_3,j_4,j_5,j_6) S_{j_4,j_5} S_{j_6,j_8} w(j_7,j_8,j_9,j_{10}) S_{j_9,j_{10}}$$
$$- 192 \sum_{j_1\cdots,j_{10}} f(j_1,j_2) S_{j_1,j_3} S_{j_2,j_7} w(j_3,j_4,j_5,j_6) S_{j_4,j_8} S_{j_5,j_9} S_{j_6,j_{10}} w(j_7,j_8,j_9,j_{10}).$$

(b) First observe that

$$W(a+b) - W(a)$$
$$= \sum_{j_1,j_2,j_3,j_4=1}^{D} w(j_1,j_2,j_3,j_4)\left[b_{j_1}b_{j_2}b_{j_3}b_{j_4} + 4b_{j_1}b_{j_2}b_{j_3}a_{j_4} + 6b_{j_1}b_{j_2}a_{j_3}a_{j_4} + 4b_{j_1}a_{j_2}a_{j_3}a_{j_4}\right]$$

is of degree at least one in b. By Problem 1.19 and Proposition 1.31a, $\int :b_{\mathrm{I}}{:}\, b_{\mathrm{J}}\, d\mu_s(b)$ vanishes unless the degree of J is at least as large as the degree of I. Hence $d^\ell/d\lambda^\ell R(\lambda)|_{\lambda=0} = 0$ for all $\ell \le 3$. Also

$$\mathrm{R}(0) = \int :0:_b F(b)\, d\mu_S(b) = 0.$$

That leaves the nontrivial cases

$$\mathrm{R}'(0) = \int :[W(a+b) - W(a)]:_b F(b)\, d\mu_S(b)$$
$$= 6 \sum_{j_1,\cdots,j_6} f(j_1,j_2) w(j_3,j_4,j_5,j_6) \int :b_{j_3}b_{j_4}a_{j_5}a_{j_6}:_b b_{j_1}b_{j_2}\, d\mu_S(b)$$
$$= -12 \sum_{j_1,\cdots,j_6} f(j_1,j_2) S_{j_1,j_3} S_{j_2,j_4} w(j_3,j_4,j_5,j_6) a_{j_5} a_{j_6}$$

and

$$\mathrm{R}''(0)$$
$$= \int :[W(a+b) - W(a)]^2:_b F(b)\, d\mu_S(b)$$
$$= 16 \sum_{j_1,\cdots,j_{10}} f(j_1,j_2) w(j_3,j_4,j_5,j_6) w(j_7,j_8,j_9,j_{10})$$
$$\times \int :b_{j_3}a_{j_4}a_{j_5}a_{j_6}b_{j_7}a_{j_8}a_{j_9}a_{j_{10}}:_b b_{j_1}b_{j_2}\, d\mu_S(b)$$
$$= 32 \sum_{j_1,\cdots,j_{10}} f(j_1,j_2) S_{j_1,j_3} S_{j_2,j_7} w(j_3,j_4,j_5,j_6) a_{j_4}a_{j_5}a_{j_6} w(j_7,j_8,j_9,j_{10}) a_{j_8}a_{j_9}a_{j_{10}}. \quad \square$$

PROBLEM 2.2. *Define, for all* $f\colon \mathcal{M}_r \to \mathbb{C}$ *and* $g\colon \mathcal{M}_s \to \mathbb{C}$ *with* $r, s \ge 1$ *and* $r+s>2$, $f \star g\colon \mathcal{M}_{R+S-2} \to \mathbb{C}$ *by*

$$f \star g(j_1,\ldots,j_{r+s-2}) = \sum_{k=1}^{D} f(j_1,\ldots,j_{r-1},k) g(k,j_r,\ldots,j_{r+s-2}).$$

Prove that

$$\|f \star g\| \le \|f\|\,\|g\| \quad and \quad |||f \star g||| \le \min\{|||f|||\,\|g\|, \|f\|\,|||g|||\}.$$

SOLUTION. By definition

$$\|f \star g\| = \max_{1\le i\le r+s-2} \max_{1\le \ell\le D} \sum_{\substack{j_1,\cdots,j_{r+s-2}=1 \\ j_i=\ell}}^{D} |f \star g(j_1,\ldots,j_{r+s-2})|.$$

We consider the case $1 \le i \le r-1$. The case $r \le i \le r+s-2$ is similar. For any $1 \le i \le r-1$,

$$\begin{aligned}
&\max_{1\le\ell\le D} \sum_{\substack{j_1,\cdots,j_{r+s-2}=1 \\ j_i=\ell}}^{D} |f \star g(j_1,\ldots,j_{r+s-2})| \\
&\le \max_{1\le\ell\le D} \sum_{\substack{j_1,\cdots,j_{r+s-2}=1 \\ j_i=\ell}}^{D} \sum_{k=1}^{D} |f(j_1,\ldots,j_{r-1},k)g(k,j_r,\ldots,j_{r+s-2})| \\
&= \max_{1\le\ell\le D} \sum_{\substack{j_1,\cdots,j_{r-1}=1 \\ j_i=\ell}}^{D} \sum_{k=1}^{D} |f(j_1,\ldots,j_{r-1},k)| \sum_{j_r,\cdots,j_{r+s-2}=1}^{D} |g(k,j_r,\ldots,j_{r+s-2})| \\
&\le \max_{1\le\ell\le D} \sum_{j_1,\cdots,j_{r-1}=1}^{D} \left[\sum_{k=1}^{D} |f(j_1,\ldots,j_{r-1},k)|\right] \max_{1\le k\le D} \sum_{j_r,\cdots,j_{r+s-2}=1}^{D} |g(k,j_r,\ldots,j_{r+s-2}| \\
&\le \max_{1\le\ell\le D} \sum_{\substack{j_1,\cdots,j_{r-1}=1 \\ j_i=\ell}}^{D} \sum_{k=1}^{D} |f(j_1,\ldots,j_{r-1},k)|\|g\| \\
&\le \|f\|\,\|g\|.
\end{aligned}$$

We prove $|||f \star g||| = |||f|||\,\|g\|$. The other case is similar.

$$\begin{aligned}
|||f \star g||| &= \sum_{j_1,\cdots,j_{r+s-2}=1}^{D} |f \star g(j_1,\ldots,j_{r+s-2})| \\
&\le \sum_{j_1,\cdots,j_{r+s-2}=1}^{D} \sum_{k=1}^{D} |f(j_1,\ldots,j_{r-1},k)g(k,j_r,\ldots,j_{r+s-2})| \\
&= \sum_{j_1,\cdots,j_{r-1}} \left[\sum_{k} |f(j_1,\ldots,j_{r-1},k)| \sum_{j_r,\cdots,j_{r+s-2}} |g(k,j_r,\ldots,j_{r+s-2})|\right] \\
&\le \sum_{j_1,\cdots,j_{r-1}} \left[\sum_{k} |f(j_1,\ldots,j_{r-1},k)|\right]\left[\max_{k} \sum_{j_r,\cdots,j_{r+s-2}} |g(k,j_r,\ldots,j_{r+s-2})|\right] \\
&\le \sum_{j_1,\cdots,j_{r-1}} \sum_{k} |f(j_1,\ldots,j_{r-1},k)|\|g\| \\
&= |||f|||\,\|g\|.
\end{aligned}$$

□

PROBLEM 2.3. *Let* $:\cdot:_S$ *denote Wick ordering with respect to the covariance* S.

(a) *Prove that if*

$$\left|\int b_{\mathrm{H}} {:}b_{\mathrm{J}}{:}_S \, d\mu_S(b)\right| \le \mathrm{F}^{|\mathrm{H}|+|\mathrm{J}|} \quad \textit{for all } \mathrm{H},\mathrm{J} \in \bigcup_{r\ge 0} \mathcal{M}_r$$

then

$$\left|\int b_{\rm H}{:}b_{\rm J}{:}_{zS}\, d\mu_{zS}(b)\right| \le (\sqrt{|z|}{\rm F})^{|{\rm H}|+|{\rm J}|} \quad \textit{for all } {\rm H},{\rm J}\in \bigcup_{r\ge 0}\mathcal{M}_r.$$

(b) *Prove that if*

$$\left|\int b_{\rm H}{:}b_{\rm J}{:}_S\, d\mu_S(b)\right| \le {\rm F}^{|{\rm H}|+|{\rm J}|} \quad \textit{and} \quad \left|\int b_{\rm H}{:}b_{\rm J}{:}_T\, d\mu_T(b)\right| \le {\rm G}^{{\rm H}|+|{\rm J}|}$$

for all H, $\rm J \in \cup_{r\ge 0}\mathcal{M}_r$, *then*

$$\left|\int b_{\rm H}{:}b_{\rm J}{:}_{S+T}\, d\mu_{S+T}(b)\right| \le ({\rm F}+{\rm G})^{|{\rm H}|+|{\rm J}|} \quad \textit{for all } {\rm H},{\rm J}\in \bigcup_{r\ge 0}\mathcal{M}_r.$$

SOLUTION. (a) For any ζ with $\zeta^2 = z$,

$$\begin{aligned}\int e^{\sum_i a_ib_i}{:}e^{\sum_i c_ib_i}{:}_{zS}\, d\mu_{zS}(b) &= \int e^{\sum_i a_ib_i}e^{\sum_i c_ib_i}e^{(\sum_{ij} zc_iS_{ij}c_j)/2}\, d\mu_{zS}(b)\\ &= e^{-(\sum_{ij} z(a_i+c_i)S_{ij}(a_j+c_j))/2}e^{1/2\sum_{ij} zc_iS_{ij}c_j}\\ &= e^{-(\sum_{ij}(\zeta a_i+\zeta c_i)S_{ij}(\zeta a_j+\zeta c_j))/2}e^{(\sum_{ij}(\zeta c_i)S_{ij}(\zeta c_j))/2}\\ &= \int e^{\sum_i \zeta a_ib_i}{:}e^{\sum_i \zeta c_ib_i}{:}_S\, d\mu_S(b).\end{aligned}$$

Differentiating with respect to a_i, $i\in{\rm H}$ and c_i, $i\in {\rm J}$ gives

$$\int b_{\rm H}{:}b_{\rm J}{:}_{zS}\, d\mu_{zS}(b) = \zeta^{{\rm H}|+|{\rm J}|}\int b_{\rm H}{:}b_{\rm J}{:}_S\, d\mu_S(b) = z^{(|{\rm H}|+|{\rm J}|)/2}\int b_{\rm H}{:}b_{\rm J}{:}_S\, d\mu_S(b).$$

Note that if $|{\rm H}|+|{\rm J}|$ is not even, the integral vanishes. So we need not worry about which square root to take.

(b) Let $f(b) = b_{\rm J}$. Then

$$\begin{aligned}&\int b_{\rm H}{:}b_{\rm J}{:}_{S+T}\, d\mu_{S+T}(b)\\ &\quad= \int (a+b)_{\rm H}{:}f{:}_{S+T}(a+b)\, d\mu_S(a)\, d\mu_T(b) && \text{by Proposition 1.21}\\ &\quad= \int (a+b)_{\rm H}{:}(a+b)_{\rm J}{:}_{\substack{a,S\\ b,T}}\, d\mu_S(a)\, d\mu_T(b) && \text{by Problem 1.21}\\ &\quad= \sum_{\substack{{\rm H}'\subset{\rm H}\\ {\rm J}'\subset{\rm J}}} \pm\int a_{{\rm H}'}{:}a_{{\rm J}'}{:}_S\, d\mu_S(a)\int b_{{\rm H}\setminus{\rm H}'}{:}b_{{\rm J}\setminus{\rm J}'}{:}_T\, d\mu_T(b).\end{aligned}$$

Hence

$$\left|\int b_{\rm H}{:}b_{\rm J}{:}_{S+T}\, d\mu_{S+T}(b)\right| \le \sum_{\substack{{\rm H}'\subset{\rm H}\\ {\rm J}'\subset{\rm J}}} F^{|{\rm H}'|+|{\rm J}'|} = G^{|{\rm H}\setminus{\rm H}'|+|{\rm J}\setminus{\rm J}'|} = ({\rm F}+{\rm G})^{|{\rm H}|+|{\rm J}|}. \quad \square$$

PROBLEM 2.4. *Let* $W(a) = \sum_{i,j=1}^D w(i,j)a_ia_j$ *with* $w(i,j) = -w(j,i)$. *Verify that*

$$W(a+b) - W(a) = \sum_{i,j=1}^D w(i,j)b_ib_j + 2\sum_{i,j=1}^D w(i,j)a_ib_j.$$

SOLUTION.

$$\begin{aligned} W(a+b)-W(a) &= \sum_{i,j=1}^{D} w(i,j)(a_i+b_i)(a_j+b_j) - \sum_{i,j=1}^{D} w(i,j)a_i,a_j \\ &= \sum_{i,j=1}^{D} w(i,j)b_ib_j + \sum_{i,j=1}^{D} w(i,j)a_ib_j + \sum_{i,j=1}^{D} w(i,j)b_ia_j. \end{aligned}$$

Now just sub in

$$\begin{aligned} \sum_{i,j=1}^{D} w(i,j)b_ia_j &= \sum_{i,j=1}^{D} -w(j,i)(-a_jb_i) \\ &= \sum_{i,j=1}^{D} w(j,i)a_jb_i = \sum_{i,j=1}^{D} w(i,j)a_ib_j. \end{aligned}$$

□

PROBLEM 2.5. *Assume Hypothesis* (HG). *Le* $s, s', m \geq 1$ *and*

$$f(b) = \sum_{\mathrm{H}\in\mathcal{M}_m} f_m(\mathrm{H})b_{\mathrm{H}}, \qquad W(b) = \sum_{\mathrm{K}\in\mathcal{M}_s} w_s(\mathrm{K})b_{\mathrm{K}}, \qquad W'(b) = \sum_{\mathrm{K}'\in\mathcal{M}_{s'}} w'_{s'}(\mathrm{K}')b_{\mathrm{K}'}.$$

(a) *Prove that*

$$\left|\int :(b):_b f(b)\, d\mu_S(b)\right| \leq m\mathrm{F}^{m+s-2} |||f_m|||\, \|S\|\, \|w_s\|.$$

(b) *Prove that* $\int :W(b)W'(b):_b f(b)\, d\mu_S(b) = 0$ *if* $m=1$ *and that, for* $m \geq 2$,

$$\left|\int :W(b)W'(b):_b f(b)\, d\mu_S(b)\right| \leq m(m-1)\mathrm{F}^{m+s+s'-4} |||f_m|||\, \|S\|^2 \|w_s\|\, \|w'_{s'}\|.$$

SOLUTION. (a) By definition

$$\begin{aligned} \int :W(b):_b f(b)\, d\mu_S(b) &= \sum_{\substack{\mathrm{H}\in\mathcal{M}_m \\ \mathrm{K}\in\mathcal{M}_s}} w_s(\mathrm{K}) f_m(\mathrm{H}) \int :b_{\mathrm{K}}:_b b_{\mathrm{H}}\, d\mu_S(b) \\ &= \sum_{\substack{\mathrm{H}\in\mathcal{M}_m \\ \widetilde{\mathrm{K}}\in\mathcal{M}_{s-1}}} \sum_{k=1}^{D} w_s\big(\widetilde{\mathrm{K}}\cdot(k)\big) f_m(\mathrm{H}) \int :b_{\widetilde{\mathrm{K}}} b_k:_b b_{\mathrm{H}}\, d\mu_S(b). \end{aligned}$$

Recall that $\widetilde{\mathrm{K}}.(k)$, the concatenation of $\widetilde{\mathrm{K}}$ and (k), is the element of $\mathcal{M}_s$ constructed by appending k to the element $\widetilde{\mathrm{K}}$ of $\mathcal{M}_{s-1}$. By Problem 1.19 (integration by parts) and Hypothesis (HG),

$$\begin{aligned} \left|\int :b_{\widetilde{\mathrm{K}}} b_k:_b b_{\mathrm{H}}\, d\mu_S(b)\right| &= \left|\sum_{\ell=1}^{D} S_{k\ell} \int :b_{\widetilde{\mathrm{K}}}: \frac{\partial}{\partial b_\ell} b_{\mathrm{H}}\, d\mu_S(b)\right| \\ &= \left|\sum_{j=1}^{m} (-1)^{j-1} S_{k,h_j} \int :b_{\widetilde{\mathrm{K}}}: b_{h_1,\ldots,\not h_j,\ldots,h_m}\, d\mu_S(b)\right| \\ &= \sum_{j=1}^{m} |S_{k,h_j}| \mathrm{F}^{s+m-2}. \end{aligned}$$

We may assume, without loss of generality, that $f_m(\mathrm{H})$ is antisymmetric under permutation of its m arguments. Hence

$$\begin{aligned}
\left|\int :W(b):_b f(b)\, d\mu_S(b)\right| &\le \sum_{\substack{\mathrm{H}\in\mathcal{M}_m\\ \widetilde{\mathrm{K}}\in\mathcal{M}_{s-1}}} \sum_{k=1}^{D}\sum_{j=1}^{m} \left|w_s\big(\widetilde{\mathrm{K}}.(k)\big)\right| |S_{k,h_j}|\, |f_m(\mathrm{H})| \mathrm{F}^{s+m-2} \\
&= m \sum_{\mathrm{H}\in\mathcal{M}_m} \sum_{k=1}^{D} \Big(\sum_{\widetilde{\mathrm{K}}\in\mathcal{M}_{s-1}} \left|w_s\big(\widetilde{\mathrm{K}}.(k)\big)\right| \Big) |S_{k,h_1}|\, |f_m(\mathrm{H})| \mathrm{F}^{s_m-2} \\
&\le m\|w_s\| \sum_{\mathrm{H}\in\mathcal{M}_m} \Big(\sum_{k=1}^{D} |S_{k,h_1}| \Big) |f_m(\mathrm{H})| \mathrm{F}^{s+m-2} \\
&\le m\|w_s\|\, \|S\| \sum_{\mathrm{H}\in\mathcal{M}_m} |f_m(\mathrm{H})| \mathrm{F}^{s+m-2} \\
&= m\|w_s\|\, \|S\|\, |||f_m||| \mathrm{F}^{s+m-2}.
\end{aligned}$$

(b) By definition

$$\begin{aligned}
&\int :W(b)W'(b):_b f(b)\, d\mu_S(b) \\
&\quad = \sum_{\substack{\mathrm{H}\in\mathcal{M}_m\\ \mathrm{K}\in\mathcal{M}_s\\ \mathrm{K}'\in\mathcal{M}_{s'}}} w_s(\mathrm{K}) w'_{s'}(\mathrm{K}') f_m(\mathrm{H}) \int :b_{\mathrm{K}} b_{\mathrm{K}'}:_b\, b_{\mathrm{H}}\, d\mu_S(b) \\
&\quad = \sum_{\substack{\mathrm{H}\in\mathcal{M}_m\\ \widetilde{\mathrm{K}}\in\mathcal{M}_{S-1}\\ \widetilde{\mathrm{K}}'\in\mathcal{M}_{s'-1}}} \sum_{k,k'=1}^{D} w_s\big(\widetilde{\mathrm{K}}.(k)\big) w_{s'}\big(\widetilde{\mathrm{K}}.(k')\big) f_m(\mathrm{H}) \int :b_{\widetilde{\mathrm{K}}} b_k b_{\widetilde{\mathrm{K}}'} b_{k'}:_b\, b_{\mathrm{H}}\, d\mu_S(b).
\end{aligned}$$

By Problem 1.19 (integration by parts), twice, and Hypothesis (HG),

$$\begin{aligned}
\left|\int :b_{\widetilde{\mathrm{K}}} b_k b_{\widetilde{\mathrm{K}}'} b_{k'}:_b b_{\mathrm{H}}\, d\mu_S(b)\right| &= \left| \sum_{\ell=1}^{D} S_{k'\ell} \int :b_{\widetilde{\mathrm{K}}} b_k b_{\widetilde{\mathrm{K}}'}: \frac{\partial}{\partial b_\ell} b_{\mathrm{H}}\, d\mu_S(b) \right| \\
&= \left| \sum_{j'=1}^{m} (-1)^{j'-1} S_{k',h_{j'}} \int :b_{\widetilde{\mathrm{K}}} b_k b_{\widetilde{\mathrm{K}}'}: b_{h_1,\dots,\not{h}_{j'},\dots,h_m}\, d\mu_S(b) \right| \\
&= \left| \sum_{j'=1}^{m} \sum_{\substack{j=1\\ j\ne j'}}^{m} \pm S_{k',h_{j'}} S_{k,h_j} \int :b_{\widetilde{\mathrm{K}}} b_{\widetilde{\mathrm{K}}'}: b_{\mathrm{H}\setminus\{h_j,h_{j'}\}}\, d\mu_S(b) \right| \\
&\le \sum_{j'=1}^{m} \sum_{\substack{j=1\\ j\ne j'}}^{m} |S_{k,h_j}| |S_{k',h_{j'}}| \mathrm{F}^{s+s'+m-4}.
\end{aligned}$$

Again, we may assume, without loss of generality, that $f_m(\mathrm{H})$ is antisymmetric under permutation of its m arguments, so that

$$\left|\int :W(b)W'(b):_b f(b)\, d\mu_S(b)\right|$$

$$\begin{aligned}
&\le \sum_{\substack{\mathrm{H}\in\mathcal{M}_m\\ \widetilde{\mathrm{K}}\in\mathcal{M}_{s-1}\\ \widetilde{\mathrm{K}}'\in\mathcal{M}_{s'-1}}} \sum_{k,k'=1}^{D}\sum_{j'=1}^{m}\sum_{\substack{j=1\\ j\neq j'}}^{m} \big|w_s(\widetilde{\mathrm{K}}.(k))\big|\,\big|w_{s'}(\widetilde{\mathrm{K}}'.(k'))\big| \\
&\qquad\qquad \times |S_{k,h_j}|\,|S_{k',h_{j'}}|\,|f_m(\mathrm{H})|F^{s+s'+m-4} \\
&= m(m-1)\sum_{\substack{\mathrm{H}\in\mathcal{M}_m\\ \widetilde{\mathrm{K}}\in\mathcal{M}_{s-1}\\ \widetilde{\mathrm{K}}'\in\mathcal{M}_{s'-1}}} \sum_{k,k'=1}^{D} \big|w_s(\widetilde{\mathrm{K}}.(k))\big|\,\big|w_{s'}(\widetilde{\mathrm{K}}'.(k'))\big| \\
&\qquad\qquad \times |S_{k,h_1}|\,|S_{k',h_2}|\,|f_m(\mathrm{H})|F^{s+s'+m-4} \\
&= m(m-1)\sum_{\mathrm{H}\in\mathcal{M}_m}\sum_{k,k'=1}^{D}\Big(\sum_{\widetilde{\mathrm{K}}}\big|w_s(\widetilde{\mathrm{K}}.(k))\big|\Big)\Big(\sum_{\widetilde{\mathrm{K}}'}\big|w_{s'}(\widetilde{\mathrm{K}}'.(k'))\big|\Big) \\
&\qquad\qquad \times |S_{k,h_1}|\,|S_{k',h_2}|\,|f_m(\mathrm{H})|\mathrm{F}^{s+s'+m-4} \\
&\le m(m-1)\|w_s\|\,\|w_{s'}\|\sum_{\mathrm{H}\in\mathcal{M}_m}\Big(\sum_{k=1}^{D}|S_{k,h_1}|\Big)\Big(\sum_{k'=1}^{D}|S_{k',h_2}|\Big)|f_m(\mathrm{H})|\mathrm{F}^{s+s'+m-4} \\
&\le m(m-1)\|w_s\|\,\|w_{s'}\|\,\|S\|^2\sum_{\mathrm{H}\in\mathcal{M}_m}|f_m(\mathrm{H})|\mathrm{F}^{s+s'+m-4} \\
&= m(m-1)\|w_s\|\,\|w_{s'}\|\,\|S\|^2|||f_m|||\mathrm{F}^{s+s'+m-4}. \qquad\square
\end{aligned}$$

PROBLEM 2.6. *Let $M > 1$. Construct a function $\nu \in C_0^\infty([M^{-2}, M^2])$ that takes values in $[0,1]$, is identically 1 on $[M^{-1/2}, M^{1/2}]$ and obeys*

$$\sum_{j=0}^{\infty}\nu(M^{2j}x) = 1$$

for $0 < x < 1$.

SOLUTION.

$$\nu_1(x) = \begin{cases} e^{-1/x^2}, & \text{if } x > 0\\ 0, & \text{if } x \le 0\end{cases}$$

is C^∞. Then

$$\nu_2(x) = \nu_1(-x)\nu_1(x+1)$$

is C^∞, strictly positive for $-1 < x < 0$ and vanishes for $x \ge 0$ and $x \le -1$,

$$\nu_3(x) = \frac{\int_{-1}^{x}\nu_2(y)\,dy}{\int_{-1}^{0}\nu_2(y)\,dy}$$

is C^∞, vanishes for $x \le -1$, is strictly increasing for $-1 < x < 0$ and identically one for $x \ge 0$ and

$$\nu_4(x) = \nu_3(1-x)\nu_1(x+1)$$

is C^∞, vanishing for $|x| \ge 2$ identically one for $|x| \le 1$ and monotone for $1 \le |x| \le 2$. Observe that, for any x, at most two terms in the sum

$$\bar\nu(x) = \sum_{m=-\infty}^{\infty}\nu_4(x+3m)$$

are nonzero, because the support of ν_4 has width $4 < 2\times 3$. Hence the sum always converges. On the other hand, as the width of the support is precisely $4 > 3$, at least one term is nonzero, so that $\bar\nu(x) > 0$. As $\bar\nu$ is periodic, it is uniformly bounded away from zero. Furthermore, for $|x| \le 1$, we have $|x + 3m| \ge 2$ for all $m \ne 0$, so that $\bar{\nu}(x) = \nu_4(x) = 1$ for all $|x| \le 1$. Hence

$$\nu_5(x) = \frac{\nu_4(x)}{\bar{\nu}(x)}$$

is C^∞, nonnegative, vanishing for $|x| \ge 2$ identically one for $|x| \le 1$ and obeys

$$\sum_{m=-\infty}^{\infty} \nu_5(x + 3m) = \sum_{m=-\infty}^{\infty} \frac{\nu_4(x + 3m)}{\bar{\nu}(x)} = 1.$$

Let $L > 0$ and set

$$\nu(x) = \nu_5\Big(\frac{1}{L} \ln x\Big).$$

Then $\nu(x)$ is C^∞ on $x > 0$, is supported on $|\ln x/L| \le 2$, i.e., for $e^{-2L} \le x \le e^{2L}$, is identically one for $|\ln x/L| \le 1$, i.e., for $e^{-L} \le x \le e^L$ and obeys

$$\sum_{m=-\infty}^{\infty} \nu(e^{3Lm}x) = \sum_{m=-\infty}^{\infty} \nu_5\Big(\frac{1}{L} \ln x + 3m\Big) = 1 \quad \text{for all } x > 0.$$

Finally, just set $M = e^{3L/2}$ and observe that $e^{2L} \le M^2$, $e^L \ge M^{1/2}$ and that when $x < 1$ only those terms with $m \ge 0$ contribute to $\sum_{m=-\infty}^{\infty} \nu(e^{3Lm}x)$. □

Problem 2.7. *Prove that*

$$\left\| \frac{\not{p} + m}{p^2 + m^2} \right\| = \frac{1}{\sqrt{p^2 + m^2}}.$$

Solution. For any matrix $\|M\|^2 = \|M^* M\|$, where M^* is the adjoint (complex conjugate of the transpose) of M. As

$$\not{p}^* = \begin{pmatrix} ip_0 & p_1 \\ -p_1 & -ip_0 \end{pmatrix}^* = \begin{pmatrix} -ip_0 & -p_1 \\ p_1 & ip_0 \end{pmatrix} = -\not{p}$$

and

$$\not{p}^2 = \begin{pmatrix} ip_0 & p_1 \\ -p_1 & -ip_0 \end{pmatrix} \begin{pmatrix} ip_0 & p_1 \\ -p_1 & -ip_0 \end{pmatrix} = \begin{pmatrix} p_0^2 + p_1^2 & 0 \\ p_0^2 + p_1^2 & 0 \end{pmatrix} = -p^2 \mathbb{1}$$

we have

$$\begin{aligned} \left\| \frac{\not{p} + m}{p^2 + m^2} \right\|^2 &= \frac{1}{(p^2 + m^2)^2} \|(\not{p}^* + m)(\not{p} + m)\| \\ &= \frac{1}{(p^2 + m^2)^2} \|(-\not{p} + m)(\not{p} + m)\| \\ &= \frac{1}{(p^2 + m^2)^2} \|-\not{p}^2 + m^2\| \\ &= \frac{1}{p^2 + m^2} \|\mathbb{1}\| = \frac{1}{p^2 + m^2}. \end{aligned}$$

□

Appendix A. Infinite-Dimensional Grassmann Algebras

PROBLEM A.1. *Let $\mathcal{I}$ be any ordered countable set and $\mathfrak{I}$ the set of all finite subsets of $\mathcal{I}$ (including the empty set). Each $\mathrm{I} \in \mathfrak{I}$ inherits an ordering from $\mathcal{I}$. Let $w\colon \mathcal{I} \to (0,\infty)$ be any strictly positive function on $\mathcal{I}$ and set*

$$W_{\mathrm{I}} = \prod_{i \in \mathrm{H}} w_i$$

with the convention $W_\varnothing = 1$. Define

$$\mathcal{V} = \ell^1(\mathcal{I}, w) = \left\{ \alpha\colon \mathcal{I} \to \mathbb{C} \;\middle|\; \sum_{i \in \mathcal{I}} w_i |\alpha_i| < \infty \right\}$$

and "the Grassmann algebra generated by $\mathcal{V}$"

$$\mathfrak{U}(\mathcal{I}, w) = \ell^1(\mathfrak{I}, W) = \left\{ \alpha\colon \mathfrak{I} \to \mathbb{C} \;\middle|\; \sum_{\mathrm{I} \in \mathfrak{I}} W_{\mathrm{I}} |\alpha_{\mathrm{I}}| < \infty \right\}.$$

The multiplication is $(\alpha\beta)_{\mathrm{I}} = \sum_{\mathrm{J} \subset \mathrm{I}} \operatorname{sgn}(\mathrm{J}, \mathrm{I} \setminus \mathrm{J}) \alpha_{\mathrm{J}} \beta_{\mathrm{I} \setminus \mathrm{J}}$ where $\operatorname{sgn}(\mathrm{J}, \mathrm{I} \setminus \mathrm{J})$ is the sign of the permutation that reorders $(\mathrm{J}, \mathrm{I} \setminus \mathrm{J})$ to I. The norm $\|\alpha\| \sum_{\mathrm{I} \in \mathfrak{I}} W_{\mathrm{I}} |\alpha_{\mathrm{I}}|$ turns $\mathfrak{U}(\mathcal{I}, w)$ into a Banach space.

(a) *Show that*

$$\|\alpha\beta\| \le \|\alpha\| \, \|\beta\|.$$

(b) *Show that if $f\colon \mathbb{C} \to \mathbb{C}$ is any function that is defined and analytic in a neighborhood of 0, then the power series $f(\alpha) = \sum_{n=0}^{\infty} f^{(n)}(0)\alpha^n/n!$ converges for all $\alpha \in \mathfrak{U}$ with $\|\alpha\|$ smaller than the radius of convergence of f.*

(c) *Prove that $\mathfrak{U}_f(\mathcal{I}) = \{\alpha\colon \mathfrak{I} \to \mathbb{C} \mid \alpha_{\mathrm{I}} = 0$ for all but finitely many $\mathrm{I}\}$ is a dense subalgebra of $\mathfrak{U}(\mathcal{I}, w)$.*

SOLUTION.

(a)
$$\begin{aligned}
\|\alpha\beta\| &= \sum_{\mathrm{I} \in \mathfrak{I}} |W_{\mathrm{I}}(\alpha\beta)_{\mathrm{I}}| = \sum_{\mathrm{I} \in \mathfrak{I}} W_{\mathrm{I}} \left| \sum_{\mathrm{J} \subset \mathrm{I}} \operatorname{sgn}(\mathrm{J}, \mathrm{I} \setminus \mathrm{J}) \alpha_{\mathrm{J}} \beta_{\mathrm{I} \setminus \mathrm{J}} \right| \\
&= \sum_{\mathrm{I} \in \mathfrak{I}} \left| \sum_{\mathrm{J} \subset \mathrm{I}} \operatorname{sgn}(\mathrm{J}, \mathrm{I} \setminus \mathrm{J}) W_{\mathrm{J}} \alpha_{\mathrm{J}} W_{rI \setminus \mathrm{J}} \beta_{\mathrm{I} \setminus \mathrm{J}} \right| \\
&\le \sum_{\mathrm{I} \in \mathfrak{I}} \sum_{\mathrm{J} \subset \mathrm{I}} W_{\mathrm{J}} |\alpha_{\mathrm{J}}| W_{\mathrm{I} \setminus \mathrm{J}} |\beta_{\mathrm{I} \setminus \mathrm{J}}| \\
&\le \sum_{\mathrm{I}, \mathrm{J} \in \mathfrak{I}} W_{\mathrm{J}} |\alpha_{\mathrm{J}}| W_{\mathrm{I}} |\beta_{rI}| = \|\alpha\| \, \|\beta\|.
\end{aligned}$$

(b)
$$\|f(\alpha)\| = \left\| \sum_{n=0}^{\infty} \frac{1}{n!} f^{(n)}(0) \alpha^n \right\| \le \sum_{n=0}^{\infty} \frac{1}{n!} |f^{(n)}(0)| \|\alpha\|^n$$

converges if $\|\alpha\|$ is strictly smaller than the radius of convergence of f.

(c) $\mathfrak{U}_f(\mathcal{I})$ is obviously closed under addition, multiplication and multiplication by complex numbers, so it suffices to prove that $\mathfrak{U}_f(\mathcal{I})$ is dense in $\mathfrak{U}(\mathcal{I}, w)$. $\mathfrak{I}$ is a countable set. Index its elements $\mathrm{I}_1, \mathrm{I}_2, \mathrm{I}_3, \dots$. Let $\alpha \in \mathfrak{U}(\mathcal{I}, w)$ and define, for each $n \in \mathbb{N}$, $\alpha^{(n)} \in \mathfrak{U}_f(\mathcal{I})$ by

$$\alpha^{(n)}_{\mathrm{I}_j} = \begin{cases} \alpha_{\mathrm{I}_j}, & \text{if } j \le n \\ 0, & \text{otherwise} \end{cases}.$$

Then

$$\lim_{n\to\infty} \|\alpha - \alpha^{(n)}\| = \lim_{n\to\infty} \sum_{j=n+1}^{\infty} W_{\mathrm{I}_j}|\alpha_{\mathrm{I}_j}| = 0$$

since $\sum_{j=1}^{\infty} W_{\mathrm{I}_j}|\alpha_{\mathrm{I}_j}|$ converges. □

PROBLEM A.2. *Let $\mathcal{I}$ be any ordered countable set and $\mathfrak{J}$ the set of all finite subsets of $\mathcal{I}$. Let*

$$\mathfrak{G} = \{\alpha \colon \mathfrak{J} \to \mathcal{C}\}$$

be the set of all *sequences indexed by $\mathfrak{J}$. Observe that our standard product $(\alpha\beta)_{\mathrm{I}} = \sum_{\mathrm{J}\subset\mathrm{I}} \operatorname{sgn}(\mathrm{J}, \mathrm{I}\setminus\mathrm{J})\alpha_{\mathrm{J}}\beta_{\mathrm{I}\setminus\mathrm{J}}$ is well-defined on $\mathfrak{G}$—for each $\mathrm{I}\in\mathfrak{J}$, $\sum_{\mathrm{J}\subset\mathrm{I}}$ is a finite sum. We now define, for each integer n, a norm on (a subset of) $\mathfrak{G}$ by*

$$\|\alpha\|_n = \sum_{\mathrm{I}\in\mathfrak{J}} 2^{n|\mathrm{I}|}|\alpha_{\mathrm{I}}|.$$

It is defined for all $\alpha \in \mathfrak{G}$ for which the series converges. Define

$$\mathfrak{U}_{\cap} = \{\alpha \in \mathfrak{G} \mid \|\alpha\|_n < \infty \text{ for all } n \in \mathbb{Z}\}$$
$$\mathfrak{U}_{\cap} = \{\alpha \in \mathfrak{G} \mid \|\alpha\|_n < \infty \text{ for some } n \in \mathbb{Z}\}.$$

(a) *Prove that if $\alpha, \beta \in \mathfrak{U}_{\cap}$ then $\alpha\beta \in \mathfrak{U}_{\cap}$.*

(b) *Prove that if $\alpha, \beta \in \mathfrak{U}_{\cup}$ then $\alpha\beta \in \mathfrak{U}_{\cup}$.*

(c) *Prove that if $f(z)$ is an entire function and $\alpha \in \mathfrak{U}_{\cap}$, then $f(a) \in \mathfrak{U}_{\cap}$.*

(d) *Prove that if $f(z)$ is analytic at the origin and $\alpha \in \mathfrak{U}_{\cup}$ has $|\alpha_{\varnothing}|$ strictly smaller than the radius of convergence of f, then $f(\alpha) \in \mathfrak{U}_{\cup}$.*

SOLUTION. (a), (b) By Problem A.1 with $w_i = 2^n$, $\|\alpha\beta\|_n \le \|\alpha\|_n\|\beta\|_n$. Part (a) follows immediately. For part (b), let $\|\alpha\|_m$, $\|\beta\|_{m'} < \infty$. Then, since $\|\alpha\|_m \le \|\alpha\|_n$ for $m \le n$,

$$\|\alpha\beta\|_{\min\{m,m'\}} \le \|\alpha\|_{\min\{m,m'\}}\|\beta\|_{\min\{m,m'\}} \le \|\alpha\|_m\|\beta\|_{m'} < \infty.$$

(c) For each $n \in \mathbb{Z}$,

$$\|f(\alpha)\|_n = \left\| \sum_{k=0}^{\infty} \frac{1}{k!} f^{(k)}(0)\alpha^k \right\|_n \le \sum_{k=0}^{\infty} \frac{1}{k!}|f^{(k)}(0)|\|\alpha\|_n^k$$

converges if $\|\alpha\|_n$ is strictly smaller than the radius of convergence of f.

(d) Let $n \in \mathbb{Z}$ be such that $\|\alpha\|_n < \infty$. Such an n exists because $\alpha \in \mathfrak{U}_{\cup}$. Observe that, for each $m \le n$,

$$\begin{aligned}\|\alpha\|_m &= \sum_{\mathrm{I}\in\mathfrak{J}} 2^{m|\mathrm{I}|}|\alpha_{\mathrm{I}}| = \sum_{\mathrm{I}\in\mathfrak{J}} 2^{(m-n)|\mathrm{I}|}2^{n|\mathrm{I}|}|\alpha_{\mathrm{I}}| \\ &\le |\alpha_{\varnothing}| + 2^{-(n-m)} \sum_{\mathrm{I}\in\mathfrak{J}} 2^{n|\mathrm{I}|}|\alpha_{\mathrm{I}}| \le |\alpha_{\varnothing}| + 2^{-(n-m)}\|\alpha\|_n.\end{aligned}$$

Consequently, for all $\alpha \in \mathfrak{U}_{\cup}$,

$$\lim_{m\to-\infty} \|\alpha\|_m = |\alpha_{\varnothing}|.$$

If $|\alpha_{\varnothing}|$ is strictly smaller than the radius of convergence of f, then, for some m, $\|\alpha\|_m$ is also strictly smaller than the radius of convergence of f and

$$\|f(\alpha)\|_m \le \sum_{k=0}^{\infty} \frac{1}{k!}|f^{(k)}(0)|\,\|\alpha\|_m^k < \infty.$$ □

PROBLEM A.3. (a) *Let $f(x) \in C^\infty(\mathbb{R}^d/L\mathbb{Z}^d)$. Define*

$$\tilde{f}_{\mathbf{p}} = \int_{\mathbb{R}^d/L\mathbb{Z}^d} d\mathbf{x}\, f(\mathbf{x}) e^{-i\langle \mathbf{p},\mathbf{x}\rangle}.$$

Prove that for every $\gamma \in \mathbb{N}$, there is a constant $C_{f,\gamma}$, such that

$$|\tilde{f}_{\mathbf{p}}| \le C_{f,\gamma} \prod_{j=1}^d (1+\mathbf{p}_j^2)^{-\gamma} \quad \textit{for all } \mathbf{p} \in \frac{2\pi}{L}\mathbb{Z}^d.$$

(b) *Let f be a distribution on $\mathbb{R}^d/L\mathbb{Z}^d$. For any $h \in C^\infty(\mathbb{R}^d/L\mathbb{Z}^d)$, set*

$$h_R(\mathbf{x}) = \frac{1}{L^d} \sum_{\substack{\mathbf{p}\in(2\pi/L)\mathbb{Z}^d \\ |\mathbf{p}|\ge R}} e^{i\langle \mathbf{p},\mathbf{x}\rangle} \tilde{h}_{\mathbf{p}}, \quad \textit{where } \tilde{h}_{\mathbf{p}} = \int_{\mathbb{R}^d/L\mathbb{Z}^d} d\mathbf{x}\, h(\mathbf{x}) e^{-i\langle \mathbf{p},\mathbf{x}\rangle}.$$

Prove that

$$\langle f, h\rangle = \lim_{R\to\infty} \langle f, h_R\rangle.$$

SOLUTION.

(a)
$$\begin{aligned}
\prod_{j=1}^d (1+\mathbf{p}_j^2)^\gamma \tilde{f}_{\mathbf{p}} &= \int_{\mathbb{R}^d/L\mathbb{Z}^d} d\mathbf{x}\, f(\mathbf{x}) \prod_{j=1}^d (1+\mathbf{p}_j^2)^\gamma e^{-i\langle \mathbf{p},\mathbf{x}\rangle} \\
&= \int_{\mathbb{R}^d/L\mathbb{Z}^d} d\mathbf{x}\, f(\mathbf{x}) \prod_{j=1}^d \left(1 - \frac{\partial^2}{\partial \mathbf{x}_j^2}\right)^\gamma e^{-i\langle \mathbf{p},\mathbf{x}\rangle} \\
&= \int_{\mathbb{R}^d/L\mathbb{Z}^d} d\mathbf{x}\, e^{-i\langle \mathbf{p},\mathbf{x}\rangle} \prod_{j=1}^d \left(1 - \frac{\partial^2}{\partial \mathbf{x}_j^2}\right)^\gamma f(\mathbf{x})
\end{aligned}$$

by integration by parts. Note that there are no boundary terms arising from the integration by parts because f is periodic. Hence

$$\prod_{j=1}^d (1+\mathbf{p}_j^2)^\gamma |\tilde{f}_{\mathbf{p}}| \le C_{f,\gamma} \equiv \int_{\mathbb{R}^d/L\mathbb{Z}^d} d\mathbf{x} \left| \prod_{j=1}^d \left(1 - \frac{\partial^2}{\partial \mathbf{x}_j^2}\right)^\gamma f(\mathbf{x}) \right| < \infty.$$

(b) By (A.2)

$$\begin{aligned}
|\langle f, h - h_R\rangle| &= \left| \left\langle f, \frac{1}{L^d} \sum_{\substack{\mathbf{p}\in(2\pi/L)\mathbb{Z}^d \\ |\mathbf{p}|\ge R}} e^{i\langle \mathbf{p},\mathbf{x}\rangle} \tilde{h}_{\mathbf{p}} \right\rangle \right| \\
&\le C_f \sup_{\mathbf{x}\in\mathbb{R}^d/L\mathbb{Z}^d} \left| \prod_{j=1}^d \left(1 - \frac{d^2}{d\mathbf{x}_j^2}\right)^{\nu_f} \left(\frac{1}{L^d} \sum_{\substack{\mathbf{p}\in(2\pi/L)\mathbb{Z}^d \\ |\mathbf{p}|\ge R}} e^{i\langle \mathbf{p},\mathbf{x}\rangle} \tilde{h}_{\mathbf{p}} \right) \right| \\
&\le \frac{C_f}{L^d} \sum_{\substack{\mathbf{p}\in(2\pi/L)\mathbb{Z}^d \\ |\mathbf{p}|\ge R}} \prod_{j=1}^d (1+\mathbf{p}_j^2)^{\nu_f} |\tilde{h}_{\mathbf{p}}|.
\end{aligned}$$

By part (a) of this problem, there is, for each $\gamma > 0$, a constant $C_{h,\gamma}$ such that

$$|\tilde{h}_{\mathbf{p}}| \le C_{h,\gamma} \prod_{j=1}^d (1+\mathbf{p}_j^2)^{-\gamma} \quad \text{for all } \mathbf{p} \in \frac{2\pi}{L}\mathbb{Z}^d.$$

Hence, choosing $\gamma = \nu_f + 1$,

$$
\begin{aligned}
|\langle f, h - h_R\rangle| &\le C_{h,\gamma}\frac{C_f}{L^d} \sum_{\substack{\mathbf{p}\in(2\pi/L)\mathbb{Z}^d\\ |\mathbf{p}|\ge R}} \prod_{j=1}^{d}(1+\mathbf{p}_j^2)^{\nu_f}(1+\mathbf{p}_j^2)^{-\gamma}\\
&= C_{h,\gamma}\frac{C_f}{L^d} \sum_{\substack{\mathbf{p}\in(2\pi/L)^d_{\mathbb{Z}}\\ |\mathbf{p}|\ge R}} \prod_{j=1}^{d}(1+\mathbf{p}_j^2)^{-1}.
\end{aligned}
$$

Since

$$
\sum_{\mathbf{p}\in(2\pi/L)\mathbb{Z}^d} \prod_{j=1}^{d}(1+\mathbf{p}_j^2)^{-1} = \Big[\sum_{\mathbf{p}_1\in(2\pi/L)\mathbb{Z}} \frac{1}{1+\mathbf{p}_1^2}\Big]^d < \infty
$$

we have

$$
\lim_{R\to\infty} \sum_{\substack{\mathbf{p}\in(2\pi/L)\mathbb{Z}^d\\ |\mathbf{p}|\ge R}} \prod_{j=1}^{d}(1+\mathbf{p}_j^2)^{-1} = 0
$$

and hence

$$
\lim_{R\to\infty} |\langle f, h - h_R\rangle| = 0. \qquad \square
$$

PROBLEM A.4. *Let f be a distribution on $\mathbb{R}^d/L\mathbb{Z}^d$. Suppose that, for each $\gamma\in\mathbb{Z}$, there is a constant $C_{f,\gamma}$ such that*

$$
|\langle f, e^{-i\langle\mathbf{p},\mathbf{x}\rangle}\rangle| \le C_{f,\gamma}\prod_{j=1}^{d}(1+\mathbf{p}_j^2)^{-\gamma} \qquad \textit{for all } \mathbf{p}\in\frac{2\pi}{L}\mathbb{Z}^d.
$$

Prove that there is a C^∞ function $F(\mathbf{x})$ on $\mathbb{R}^d/L\mathbb{Z}^d$ such that

$$
\langle f, h\rangle = \int_{\mathbb{R}^d/L\mathbb{Z}^d} d\mathbf{x}\, F(\mathbf{x})h(\mathbf{x}).
$$

PROOF. Define $\tilde F_{\mathbf{p}} = \langle f, e^{-i\langle\mathbf{p},\mathbf{x}\rangle}\rangle$ and $F(\mathbf{x}) = L^{-d}\sum_{\mathbf{q}\in(2\pi/L)\mathbb{Z}^d} e^{i\langle\mathbf{q},\mathbf{x}\rangle}\tilde F_{\mathbf{q}}$. By hypothesis $|\tilde F_{\mathbf{q}}| \le C_{f,\gamma}\prod_{j=1}^{d}(1+\mathbf{q}_j^2)^{-\gamma}$ for all $\mathbf{q}\in(2\pi/L)\mathbb{Z}^d$ and $\gamma\in\mathbb{N}$. Consequently, all termwise derivatives of the series defining F converge absolutely and uniformly in $\mathbf{x}$. Hence $F\in C^\infty(\mathbb{R}^d/L\mathbb{Z}^d)$. Define the distribution φ by

$$
\langle \varphi, h\rangle = \int_{\mathbb{R}^d/L\mathbb{Z}^d} d\mathbf{x}\, F(\mathbf{x})h(\mathbf{x}).
$$

We wish to show that $\varphi = f$. As

$$
\begin{aligned}
\langle\varphi, e^{-i\langle\mathbf{p},\mathbf{x}\rangle}\rangle &= \int_{\mathbb{R}^d/L\mathbb{Z}^d} d\mathbf{x}\, F(\mathbf{x})e^{-i\langle\mathbf{p},\mathbf{x}\rangle} = \frac{1}{L^d}\sum_{\mathbf{q}\in(2\pi/L)\mathbb{Z}^d}\tilde F_{\mathbf{q}}\int_{\mathbb{R}^d/L\mathbb{Z}^d} d\mathbf{x}\, e^{i\langle(\mathbf{q}-\mathbf{p}),\mathbf{x}\rangle}\\
&= \frac{1}{L^d}\sum_{\mathbf{q}\in(2\pi/L)\mathbb{Z}^d}\tilde F_{\mathbf{q}}L^d\delta_{\mathbf{p},\mathbf{q}} = \tilde F_{\mathbf{p}} = \langle f, e^{-i\langle\mathbf{p},\mathbf{x}\rangle}\rangle
\end{aligned}
$$

we have that $\langle\varphi, h\rangle = \langle f, h\rangle$ for any h that is a finite linear combination of $e^{-i\langle\mathbf{p},\mathbf{x}\rangle}$, $\mathbf{p}\in(2\pi/L)\mathbb{Z}^d$. It now suffices to apply part (b) of Problem A.3. $\square$

PROBLEM A.5. *Prove that*

(a) $\mathbb{A}\mathbb{A}^\dagger f = \mathbb{A}^\dagger\mathbb{A} f + f$

(b) $\mathbb{A}h_0 = 0$

(c) $\mathbb{A}^\dagger \mathbb{A} h_\ell = \ell h_\ell$
(d) $\langle h_\ell, h_{\ell'} \rangle = \delta_{\ell,\ell'}$. *Here* $\langle f, g \rangle = \int \bar{f}(x) g(x)\, dx$.
(e) $x^2 - d^2/dx^2 = 2\mathbb{A}^\dagger \mathbb{A} + 1$.

SOLUTION. (a) By definition and the fact that $d/dx(xf) = f + x\,df/dx$,

$$\begin{aligned}
\mathbb{A}\mathbb{A}^\dagger &= \frac{1}{2}\left(x + \frac{d}{dx}\right)\left(x - \frac{d}{dx}\right) f \\
&= \frac{1}{2}\left(x^2 + \frac{d}{dx}x - x\frac{d}{dx} - \frac{d^2}{dx^2}\right) f = \frac{1}{2}\left(x^2 + 1 - \frac{d^2}{dx^2}\right) f \\
\mathbb{A}^\dagger \mathbb{A} f &= \frac{1}{2}\left(x - \frac{d}{dx}\right)\left(x + \frac{d}{dx}\right) f \\
&= \frac{1}{2}\left(x^2 - \frac{d}{dx}x + x\frac{d}{dx} - \frac{d^2}{dx^2}\right) f = \frac{1}{2}\left(x^2 - 1 - \frac{d^2}{dx^2}\right) f.
\end{aligned} \tag{D.3}$$

Subtracting

$$\mathbb{A}\mathbb{A}^\dagger f - \mathbb{A}^\dagger \mathbb{A} f = f.$$

(b) $$\sqrt{2}\pi^{1/4}\mathbb{A} h_0 = \left(x + \frac{d}{dx}\right) e^{-1/2x^2} = x e^{-1/2x^2} - x e^{-1/2x^2} = 0.$$

(c) The proof is by induction on ℓ. When $\ell = 0$,

$$\mathbb{A}^\dagger \mathbb{A} h_0 = 0 = 0 h_0$$

by part (b). If $\mathbb{A}^\dagger \mathbb{A} h_\ell = \ell h_\ell$, then, by part (a),

$$\begin{aligned}
\mathbb{A}^\dagger \mathbb{A} h_{\ell+1} &= \mathbb{A}^\dagger \mathbb{A} \frac{1}{\sqrt{\ell+1}} \mathbb{A}^\dagger h_\ell = \frac{1}{\sqrt{\ell+1}} \mathbb{A}^\dagger \mathbb{A}^\dagger \mathbb{A} h_\ell + \frac{1}{\sqrt{\ell+1}} \mathbb{A}^\dagger h_\ell \\
&= \frac{1}{\sqrt{\ell+1}} \mathbb{A}^\dagger \ell h_\ell + \frac{1}{\sqrt{\ell+1}} \mathbb{A}^\dagger h_\ell = (\ell+1)\frac{1}{\sqrt{\ell+1}} \mathbb{A}^\dagger h_\ell = (\ell+1) h_{\ell+1}.
\end{aligned}$$

(d) By part (c)

$$\begin{aligned}
(\ell - \ell')\langle h_\ell, h_{\ell'} \rangle = \langle \ell h_\ell, h_{\ell'} \rangle - \langle h_\ell, \ell' h_{\ell'} \rangle &= \langle \mathbb{A}^\dagger \mathbb{A} h_\ell, h_{\ell'} \rangle - \langle h_\ell, \mathbb{A}^\dagger \mathbb{A} h_{\ell'} \rangle \\
&= \langle h_\ell, \mathbb{A}^\dagger \mathbb{A} h_{\ell'} \rangle - \langle h_\ell, \mathbb{A}^\dagger \mathbb{A} h_{\ell'} \rangle = 0
\end{aligned}$$

since $\mathbb{A}^\dagger$ and $\mathbb{A}$ are adjoints. So, if $\ell \neq \ell'$, then $\langle h_\ell, h_{\ell'} \rangle = 0$. We prove that $\langle h_\ell, h_\ell \rangle = 1$, by induction on ℓ. For $\ell = 0$

$$\langle h_0, h_0 \rangle = \frac{1}{\sqrt{\pi}} \int e^{-x^2}\, dx = 1.$$

If $\langle h_\ell, h_\ell \rangle = 1$, then

$$\begin{aligned}
\langle h_{\ell+1}, h_{\ell+1} \rangle &= \frac{1}{\ell+1} \langle \mathbb{A}^\dagger h_\ell, \mathbb{A}^\dagger h_\ell \rangle = \frac{1}{\ell+1} \langle h_\ell, \mathbb{A}\mathbb{A}^\dagger h_\ell \rangle \\
&= \frac{1}{\ell+1} \langle h_\ell, (\mathbb{A}^\dagger \mathbb{A} + 1) h_\ell \rangle = \frac{1}{\ell+1} \langle h_\ell, (\ell+1) h_\ell \rangle = 1.
\end{aligned}$$

(e) Adding the two rows of (D.3) gives

$$\mathbb{A}\mathbb{A}^\dagger f + \mathbb{A}^\dagger \mathbb{A} f = \left(x^2 - \frac{d^2}{dx^2}\right) f$$

or

$$x^2 - \frac{d^2}{dx^2} = \mathbb{A}\mathbb{A}^\dagger + \mathbb{A}^\dagger \mathbb{A}.$$

Part (a) may be expressed $\mathbb{A}\mathbb{A}^\dagger = \mathbb{A}^\dagger\mathbb{A} + 1$. Subbing this in gives part (e). □

PROBLEM A.6. *Show that the constant function $f = 1$ is in $\mathcal{V}_\gamma$ for all $\gamma < -1$.*

SOLUTION. We first show that, under the Fourier transform convention,

$$\hat{g}(k) = \frac{1}{\sqrt{2\pi}} \int g(x) e^{-ikx}\, dx$$

$\hat{h}_\ell = (-i)^\ell h_\ell$. The Fourier transform of h_0 is

$$\begin{aligned}\hat{h}_0(k) &= \frac{1}{\pi^{1/4}} \frac{1}{\sqrt{2\pi}} \int e^{-x^2/2} e^{-ikx}\, dx = \frac{1}{\pi^{1/4}} \frac{1}{\sqrt{2\pi}} \int e^{-(x+ik)^2/2} e^{-k^2/2}\, dx \\ &= \frac{1}{\pi^{1/4}} e^{-1/2k^2} \frac{1}{\sqrt{2\pi}} \int e^{-x^2/2}\, dx = \frac{1}{\pi^{1/4}} e^{-k^2/2} = h_0(k)\end{aligned}$$

as desired. Furthermore

$$\begin{aligned}\widehat{\frac{dg}{dx}}(k) &= \frac{1}{\sqrt{2\pi}} \int g'(x) e^{-ikx}\, dx = -\frac{1}{\sqrt{2\pi}} \int g(x) \frac{d}{dx} e^{-ikx}\, dx = ik\hat{g}(k) \\ \widehat{xg}(k) &= \frac{1}{\sqrt{2\pi}} \int xg(x) e^{-ikx}\, dx = i\frac{d}{dk} \frac{1}{\sqrt{2\pi}} \int g(x) e^{-ikx}\, dx = i\frac{d}{dk}\hat{g}(k)\end{aligned}$$

so that

$$\widehat{\mathbb{A}^\dagger g} = \frac{1}{\sqrt{2}} \left(\left(x - \frac{d}{dx} \right) g \right)^{\widehat{}} = \frac{1}{\sqrt{2}} \left(i\frac{d}{dk} - ik \right) \hat{g} = (-i)\mathbb{A}^\dagger \hat{g}.$$

As $h_\ell = 1/\sqrt{\ell}\mathbb{A}^\dagger h_{\ell-1}$, the claim $\hat{h}_\ell = (-i)^\ell h_\ell$ follows easily by induction. Now let f be the function that is identically one. Then

$$\breve{f}_\ell = \langle f, h_\ell \rangle = \int h_\ell(x)\, dx = \sqrt{2\pi}\hat{h}_\ell(0) = \sqrt{2\pi}(-i)^\ell h_\ell(0)$$

is bounded uniformly in ℓ [**AS**, p. 287]. Since $\sum_{i\in\mathbb{N}}(1+2i)^\gamma$ converges for all $\gamma < -1$, f is in $\mathcal{V}_\gamma$ for all $\gamma < -1$. □

Appendix B. Pfaffians

PROBLEM B.1. *Let $T = (T_{ij})$ be a complex $n \times n$ matrix with $n = 2m$ even and let $S = \frac{1}{2}(T - T^t)$ be its skew symmetric part. Prove that* $\operatorname{Pf} T = \operatorname{Pf} S$.

SOLUTION. Define, for any n matrices, $T^{(k)} = (T^{(k)}_{ij})$, $1 \le k \le n$ each of size $n \times n$,

$$\operatorname{pf}(T^{(1)}, T^{(2)}, \ldots, T^{(n)}) = \frac{1}{2^m m!} \sum_{i_n, \ldots, i_n = 1}^{n} \varepsilon^{i_1 \ldots i_n} T^{(1)}_{i_1 i_2} \cdots T^{(n)}_{i_{n-1} i_n}.$$

Then

$$\begin{aligned}\operatorname{pf}(T^{(1)^t}, T^{(2)}, \ldots, T^{(n)}) &= \frac{1}{2^m m!} \sum_{i_n, \ldots, i_n = 1}^{n} \varepsilon^{i_1 \ldots i_n} T^{(1)^t}_{i_1 i_2} T^{(2)}_{i_3 i_4} \cdots T^{(n)}_{i_{n-1} i_n} \\ &= \frac{1}{2^m m!} \sum_{i_1, \ldots, i_n = 1}^{n} \varepsilon^{i_1 \ldots i_n} T^{(1)}_{i_2 i_1} T^{(2)}_{i_3 i_4} \cdots T^{(n)}_{i_{n-1} i_n} \\ &= \frac{1}{2^m m!} \sum_{j_1, j_2, i_3, \ldots, i_n = 1}^{n} \varepsilon^{j_2, j_1, i_3 \ldots i_n} T^{(1)}_{j_1 j_2} T^{(2)}_{i_3 i_4} \cdots T^{(n)}_{i_{n-1} i_n}\end{aligned}$$

$$= -\frac{1}{2^m m!} \sum_{j_1,j_2,i_3,\dots,i_n=1}^{n} \varepsilon^{j_1,j_2,i_3,\dots,i_n} T^{(1)}_{j_1 j_2} T^{(2)}_{i_3 i_4} \cdots T^{(n)}_{i_{n-1} i_n}$$
$$= -\operatorname{pf}(T^{(1)}, T^{(2)}, \dots, T^{(n)}).$$

Because $\operatorname{pf}(T^{(1)}, T^{(2)}, \dots, T^{(n)})$ is linear in $T^{(1)}$,

$$\operatorname{pf}\left(\frac{1}{2}\{T^{(1)} - T^{(1)^t}\}, T^{(2)}, \dots, T^{(n)}\right)$$
$$= \frac{1}{2}\operatorname{pf}(T^{(1)}, T^{(2)}, \dots, T^{(n)}) - \frac{1}{2}\operatorname{pf}(T^{(1)^t}, T^{(2)}, \dots, T^{(n)})$$
$$= \frac{1}{2}\operatorname{pf}(T^{(1)}, T^{(2)}, \dots, T^{(n)}) + \frac{1}{2}\operatorname{pf}(T^{(1)}, T^{(2)}, \dots, T^{(n)})$$
$$= \operatorname{pf}(T^{(1)}, T^{(2)}, \dots, T^{(n)}).$$

Applying the same reasoning to the other arguments of $\operatorname{pf}(T^{(1)}, T^{(2)}, \dots, T^{(n)})$.

$$\operatorname{pf}\left(\frac{1}{2}\{T^{(1)} - T^{(1)^t}\}, \dots, \frac{1}{2}\{T^{(n)} - T^{(n)^t}\}\right) = \operatorname{pf}(T^{(1)}, T^{(2)}, \dots, T^{(n)}).$$

Setting $T^{(1)} = T^{(2)} = \cdots = T^{(n)} = T$ gives the desired result. □

PROBLEM B.2. *Let* $S = \left(\begin{smallmatrix} 0 & S_{12} \\ S_{21} & 0 \end{smallmatrix}\right)$ *with* $S_{21} = -S_{12} \in \mathbb{C}$. *Show that* $\operatorname{Pf} S = S_{12}$.

SOLUTION. By (B.1) with $m = 1$,

$$\operatorname{Pf}\begin{pmatrix} 0 & S_{12} \\ S_{21} & 0 \end{pmatrix} = \frac{1}{2}\sum_{1\le k,l\le 2} \varepsilon^{kl} S_{kl} = \frac{1}{2}(S_{12} - S21) = S_{12}.$$ □

PROBLEM B.3. *Let* $\alpha_1, \dots, \alpha_r$ *be complex numbers and let* S *be the* $2r \times 2r$ *skew symmetric matrix*

$$S = \bigoplus_{m=1}^{r} \begin{bmatrix} 0 & \alpha_m \\ -\alpha_m & 0 \end{bmatrix}.$$

Prove that $\operatorname{Pf}(S) = \alpha_1 \alpha_2 \dots \alpha_r$.

SOLUTION. All matrix elements of S are zero, except for r 2×2 blocks running down the diagonal. For example, if $r = 2$,

$$S = \begin{bmatrix} 0 & \alpha_1 & 0 & 0 \\ -\alpha_1 & 0 & 0 & 0 \\ 0 & 0 & 0 & \alpha_2 \\ 0 & 0 & -\alpha_2 & 0 \end{bmatrix}.$$

By (B.1″),

$$\operatorname{Pf}(S) = \sum_{\mathbf{P}\in\mathcal{P}_r^<} \varepsilon^{k_1\ell_1\dots k_r\ell_r} S_{k_1\ell_1} \dots S_{k_r\ell_r}$$
$$= \sum_{\substack{1\le k_1<k_2<\cdots<k_r\le 2r \\ 1\le k_i<\ell_i\le 2r, 1\le i\le r}} \varepsilon^{k_1\ell_1\dots k_r\ell_r} S_{k_1,\ell_1} \dots S_{k_r\ell_r}.$$

The conditions $1 \le k_1 < k_2 < \cdots < k_r \le 2r$ and $1 \le k_i < \ell_i \le 2r$, $1 \le i \le r$ combined with the requirement that the $k_1, \ell_1, \dots, k_r, \ell_r$ all be distinct force $k_1 = 1$. Then $S_{k_1\ell 1}$ is nonzero only if $\ell_1 = 2$. When $k_1 = 1$ and $\ell_2 = 2$, the conditions $k_1 < k_2 < \cdots < k_r \le 2r$ and $1 \le k_i < \ell_i \le 2r$, $1 \le i \le r$ combined with the requirement that the $k_1, \ell_1, \ldots, k_r, \ell_r$ all be distinct force $k_2 = 3$. Then $S_{k_2\ell_2}$ is nonzero only if $\ell_2 = 4$. Continuing in this way

$$\mathrm{Pf}(S) = \varepsilon^{1,2,\ldots(2r-1),2r} S_{1,2} \cdots S_{2r-1,2r} = \alpha_1\alpha_2 \cdots \alpha_r. \qquad \square$$

PROBLEM B.4. *Let S be a skew symmetric $D \times D$ matrix with D odd. Prove that* $\det S = 0$.

SOLUTION. As $S^t = -S$,

$$\det S = \det S^t = \det(-S) = (-1)^D \det S.$$

As D is odd, $\det S = -\det S$ and $\det S = 0$. $\square$

Appendix C. Propagator Bounds

PROBLEM C.1. *Prove, under the hypotheses of Proposition* C.1, *that there are constants $C, C' > 0$, such that if $|\mathbf{k} - \mathbf{k}'_c| \le C$, then there is a point $\mathbf{p}' \in \mathcal{F}$ with $(\mathbf{k} - \mathbf{p}') \cdot \hat{\mathbf{t}} = 0$ and $|\mathbf{k} - \mathbf{p}'| \le C'|e(\mathbf{k})|$.*

SOLUTION. We are assuming that ∇e does not vanish on $\mathcal{F}$. As $\nabla e(\mathbf{k}'_c) \| \hat{\mathbf{n}}$, $\nabla e(\mathbf{k}'_c) \cdot \hat{\mathbf{n}}$ is nonzero. Hence $\nabla e(\mathbf{k}) \cdot \hat{\mathbf{n}}$ is bounded away from zero, say by C_2, on a neighborhood of $\mathbf{k}'_c$. This neighborhood contains a square centered on $\mathbf{k}'_c$ with sides parallel to $\hat{\mathbf{n}}$ and $\hat{\mathbf{t}}$, such that $\mathcal{F}$ is a graph over the sides parallel to $\hat{\mathbf{t}}$. Pick C

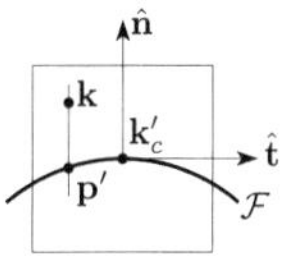

to be half the length of the sides of this square. Then any $\mathbf{k}$ within a distance C of $\mathbf{k}'_c$ is also in this square. For each $\mathbf{k}$ in this square there is a ε such that $\mathbf{k} + \varepsilon\hat{\mathbf{n}} \in \mathcal{F}$ and the line joining $\mathbf{k}$ to $\mathbf{k} + \varepsilon\hat{\mathbf{n}} \in \mathcal{F}$ also lie in the square. Then

$$\begin{aligned} |e(\mathbf{k})| = |e(\mathbf{k}) - e(\mathbf{k} + \varepsilon\hat{\mathbf{n}})| &= \left| \int_0^\varepsilon dt \frac{d}{dt} e(\mathbf{k} + t\hat{\mathbf{n}}) \right| \\ &= \left| \int_0^\varepsilon dt \nabla e(\mathbf{k} + t\hat{\mathbf{n}}) \cdot \hat{\mathbf{n}} \right| \ge C_2|\varepsilon| = C_2|\mathbf{k} - (\mathbf{k} + \varepsilon\hat{\mathbf{n}})|. \end{aligned}$$

Pick $\mathbf{p}' = \mathbf{k} + \varepsilon\hat{\mathbf{n}}$. $\square$

PROBLEM C.2. *Let $f\colon \mathbb{R}^d \to \mathbb{R}$ and $g\colon \mathbb{R} \to \mathbb{R}$. Let $1 \le i_1, \ldots, i_n \le d$. Prove that*

$$\Big(\prod_{\ell=1}^n \frac{\partial}{\partial x_{i_\ell}} \Big) g\big(f(x)\big) = \sum_{m=1}^n \sum_{(I_1,\ldots,I_m)\in\mathcal{P}_m^{(n)}} g^{(m)}\big(f(x)\big) \prod_{p=1}^m \prod_{\ell \in I_p} \frac{\partial}{\partial x_{i_\ell}} f(x)$$

where $\mathcal{P}_m^{(n)}$ is the set of all partitions of $(1, \ldots, n)$ into m nonempty subsets $I_1, \ldots,$ I_m with, for all $i < i'$, the smallest element of I_i smaller than the smallest element of $I_{i'}$.

SOLUTION. the proof is easy by induction on n. When $n = 1$,

$$\frac{\partial}{\partial x_{i_1}} g\big(f(x)\big) = g'\big(f(x)\big) \frac{\partial}{\partial x_{i_1}} f(x)$$

which agrees with the desired conclusion, since $\mathcal{P}_1^{(1)} = \{(1)\}$, m must be one and I_1 must be (1). Once the result is known for $n-1$,

$$
\begin{aligned}
&\left(\prod_{\ell=1}^{n} \frac{\partial}{\partial x_{i_\ell}}\right) g(f(x)) \\
&= \frac{\partial}{\partial x_{i_n}} \sum_{m=1}^{m-1} \sum_{(I_1,\ldots,I_m)\in\mathcal{P}_m^{(n-1)}} g^{(m)}(f(x)) \prod_{p=1}^{m} \prod_{\ell\in I_p} \frac{\partial}{\partial x_{i_\ell}} f(x) \\
&= \sum_{m=1}^{n-1} \sum_{(I_1,\ldots,I_m)\in\mathcal{P}_m^{(n)}} g^{(m+1)}(f(x)) \frac{\partial}{\partial x_{i_n}} f(x) \prod_{p=1}^{m} \prod_{\ell\in I_p} \frac{\partial}{\partial x_{i_\ell}} f(x) \\
&\quad + \sum_{m=1}^{n-1} \sum_{(I_1,\ldots,I_m)\in\mathcal{P}_m^{(n)}} \sum_{q=1}^{m} g^{(m)}(f(x)) \left(\prod_{\substack{p=1\\p\neq q}}^{m} \frac{\partial}{\partial x_{i_\ell}} f(x)\right)\left(\frac{\partial}{\partial x_{i_n}} \prod_{\ell\in I_q} \frac{\partial}{\partial x_{i_\ell}} f(x)\right) \\
&= \sum_{m'=1}^{n} \sum_{(I'_1,\ldots,I'_{m'})\in\mathcal{P}_{m'}^{(n)}} g^{(m')}(f(x)) \prod_{p=1}^{m'} \prod_{\ell\in I'_p} \frac{\partial}{\partial x_{i_\ell}} f(x).
\end{aligned}
$$

For the first sum after the second equals sign, we made the changes of variables $m' = m+1$, $I'_1 = I_1$, $\ldots$, $I'_{m'-1} = I_m$, $I'_{m'} = (n)$. They account for all terms of $(I'_1,\ldots,I'_{m'}) \in \mathcal{P}_{m'}^{(n)}$ for which n is the only element of some I'_ℓ. For each $1 \le q \le m$, the qth term in the second sum after the second equals sign accounts for all terms of $(I'_1,\ldots,I'_{m'}) \in \mathcal{P}_{m'}^{(n)}$ for which n appears in I'_q and is not the only element of I'_q. $\square$

Bibliography

[AR] A. Abdesselam and V. Rivasseau, *Explicit fermionic tree expansions*, Lett. Math. Phys., **41** (1998), 77–88.

[AS] M. Abramowitz and I. A. Stegun (eds.), *Handbook of mathematical functions with formulas, graphs, and mathematical tables*, Dover, New York, 1966.

[Be] F. A. Berezin, *The method of second quantization*, Pure Appl. Phys., vol. 24, Academic Press, New York-London, 1966.

[BS] F. Berezin and M. Shubin, *The Schrödinger equation*, Math. Appl. (Soviet Ser.), vol. 66, Kluwer, Dordrecht, 1991, Supplement 3: D. Leĭtes, Quantization and supermanifolds.

[FMRS] J. Feldman, J. Magnen, V. Rivasseau, and R. Sénéor, *A renormalizable field theory: the massive Gross-Neveu model in two dimensions*, Comm. Math. Phys. **103** (1986), 67–103.

[FST] J. Feldman, M. Salmhofer, and E. Trubowitz, *Perturbation theory around nonnested Fermi surfaces. I. Keeping the Fermi surface fixed*, J. Statist. Phys. **84** (1996), 1209–1336.

[GK] K. Gawedzki and A. Kupiainen, *Gross-Neveu model through convergent perturbation expansions*, Comm. Math. Phys. **102** (1986), 1–30.

[RS] M. Reed and B. Simon, *Methods of modern mathematical physics. I. Functional analysis*, Academic Press, New York-London, 1972.

The "abstract" fermionic expansion discussed in Chapter 2 is similar to that in

[FKT] J. Feldman, H. Knörrer, and E. Trubowitz, *A nonperturbative representation for fermionic correlation functions*, Comm. Math. Phys., **195** (1998), 465–493.

[FMRT] J. Feldman, J. Magnen, V. Rivasseau, and E. Trubowitz, *An infinite volume expansion for many fermion Green's functions*, Helv. Phys. Acta **65** (1992), 679–721.

The latter also contains a discussion of sectors. Both are available over the web at `http://www.math.ubc.ca/~feldman/research.html`.

Selected Titles in This Series

16 **Joel Feldman, Horst Knörrer, and Eugene Trubowitz,** Fermionic functional integrals and the renormalization group, 2002
15 **José I. Burgos Gil,** The regulators of Beilinson and Borel, 2002
14 **Eyal Z. Goren,** Lectures on Hilbert modular varieties and modular forms, 2002
13 **Michael Baake and Robert V. Moody, Editors,** Directions in mathematical quasicrystals, 2000
12 **Masayoshi Miyanishi,** Open algebraic surfaces, 2001
11 **Spencer J. Bloch,** Higher regulators, algebraic K-theory, and zeta functions of elliptic curves, 2000
10 **James D. Lewis,** A survey of the Hodge conjecture, Second Edition, 1999
9 **Yves Meyer,** Wavelets, vibrations and scaling, 1998
8 **Ioannis Karatzas,** Lectures on the mathematics of finance, 1996
7 **John Milton,** Dynamics of small neural populations, 1996
6 **Eugene B. Dynkin,** An introduction to branching measure-valued processes, 1994
5 **Andrew Bruckner,** Differentiation of real functions, 1994
4 **David Ruelle,** Dynamical zeta functions for piecewise monotone maps of the interval, 1994
3 **V. Kumar Murty,** Introduction to Abelian varieties, 1993
2 **M. Ya. Antimirov, A. A. Kolyshkin, and Rémi Vaillancourt,** Applied integral transforms, 1993
1 **D. V. Voiculescu, K. J. Dykema, and A. Nica,** Free random variables, 1992